Thinking Through Math Word Problems

Strategies for Intermediate
Elementary School Students

Thinking Through Math Word Problems

Arthur Whimbey ■ Jack Lochhead ■ Paula Potter

1990

LAWRENCE ERLBAUM ASSOCIATES, PUBLISHERS
Hillsdale, New Jersey   Hove and London

Copyright © 1990 by Lawrence Erlbaum Associates, Inc.
All rights reserved. No part of this book may be reproduced in
any form, by photostat, microfilm, retrieval system, or any other
means, without the prior written permission of the publisher.

Lawrence Erlbaum Associates, Inc., Publisher
365 Broadway
Hillsdale, New Jersey 07642

Printed in the United States of America
10 9 8 7 6 5 4 3 2 1

Contents

1 Addition & Subtraction *1*

2 Multiplication *37*

3 Division *71*

4 Fractions *109*

1 Addition & Subtraction

Lesson 1

1. This is a heart-shaped box with 4 cookies.

   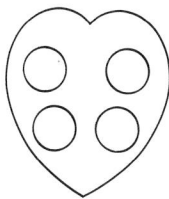

   Draw a heart-shaped box with 6 cookies in it.

2. Here is a round cage with 2 dogs and a square cage with 3 snakes.

    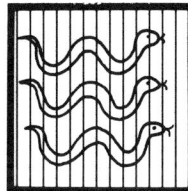

   Draw a round cage with 1 dog and a square cage with 4 snakes.

3. A heart-shaped box has 3 snakes and a square box has 8 snakes.

   Draw the heart-shaped box and show the snakes in it.

   Draw the square box and show the snakes in it.

   How many snakes are there altogether?

4. Here is a tank with 2 fish.

   A pet store has a small tank with 3 fish and a large tank with 9 fish.

   Draw the tanks with the fish in them.

   How many fish are there altogether?

2 ADDITION & SUBTRACTION

5. Here is a round cage with 2 kittens.

Al has a round cage with 1 kitten, a larger round cage with 2 kittens, and a square cage with 4 kittens.

Draw the cages and show the kittens in them.

How many kittens are there in the 2 round cages together?____

How many kittens are there in all 3 cages?____

6. This is a box with 2 fish eyeballs.

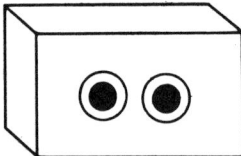

Monster Max has 3 boxes. Each box has 5 fish eyeballs.

Draw all the boxes and show the eyeballs in them.

How many eyeballs are there altogether?

7. This is a box with 2 snakes.

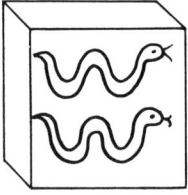

Tarzan has 3 small boxes and 2 large boxes.
Each small box has 4 snakes and each large box has 7 snakes.

Draw all the boxes and show the snakes in them.

How many total snakes are there in the small boxes?____

How many total snakes are there in the large boxes?____

How many total snakes does Tarzan have?____

8. This is a box with 2 teddy bears?

   In the store each small box has 1 bear and each large box has 3 bears. There are 4 small boxes and 2 large boxes.

   Draw all the boxes and show the bears in them.

   How many total bears are there in the small boxes?____

   How many total bears are there in the large boxes?____

   How many bears are there altogether?

9. Mad Max saw 4 coats in the closet. He put 2 spiders in each coat. Then he saw 3 hats. He put 5 spiders in each hat.

   How many total spiders did he put in the clothing?____

10. Mr. Frog is just learning to talk. For 3 days he said 2 words each day. For the next 4 days he said 5 words each day.

    How many words did he say in all 7 days?____

Lesson 2

1. There were 6 candies in a box. Paul ate 2 candies. The candies Paul ate are crossed out.

   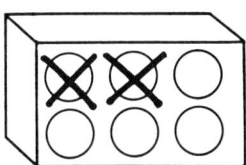

   How many candies are left?____

2. Draw a box with 8 cookies in it.
   Pretend you ate 3 cookies.

   Cross out the 3 cookies you ate.

   How many cookies are left?____

4   ADDITION & SUBTRACTION

3. Draw a box with 10 worms like this.

   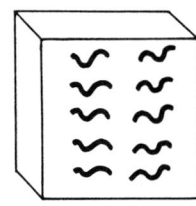

   Bill ate 3 worms and Cynthia ate 4.

   Cross out the worms the children ate from your box.

   How many worms are left in the box?____

4. Rafael had 18 fish in his tank. He gave 5 fish to his cousin.

   How many fish did Rafael have left?____

5. Peter had 25 spiders. He gave 4 spiders to Addison and 10 to Mike.

   How many spiders did Peter have left?____

6. There are 20 children in class. 5 children left their math books home. 3 other children lost their math books.

   How many children have their math books in class?

7. Tom had $10. He spent $4 for a turtle.
   How much did he have left?

8. Keithon had $9. He bought a birthday cake for $6 and candles for $1.
   How much did he have left?

9. Paula bought a camera for $14. She gave the clerk $20.
   How much change should she receive back?

10. Morris bought a snake for $14 and a spider for $8. He gave the clerk $30.
    How much change did he get?

11. There were 150 ants in a jar. Alvin thought they were raisins and ate 12. Betty saw Alvin eating them, so she ate 14 herself.
    How many ants were left in the jar?

Lesson 3

1. Benny Bear woke up thirsty. He drank a can of soda. Then he drank 16 more cans of soda.

   How many cans of soda did he drink altogether?

2. Godzilla was hungry. He went to Burger King and ate the building. Then he ate 5 more Burger King buildings and 3 McDonald's buildings.

   How many buildings did he eat altogether?____

3. There were 17 trees in a field. It started to rain and 8 trees ran away.

   How many trees were left?____

4. Tim had a can with 75 worms. He put 23 worms in his brother's bed and 17 worms in his sister's bed.

   How many worms were left?____

5. Jack bought a shirt for $8 and a belt for $3. He gave the clerk a $20 bill.

   How much change did he get?____

6. Cheryl bought a monkey for $300 and a puppy for $100. She gave the clerk a check for $500.

   How much change should she receive?____

7. Mrs. Gold took her class of 25 children to the zoo. The lions ate 3 children and the bears ate 5.

   How many children were left?____

8. A box had 20 roaches and 30 spiders. Bill thought they were candies. He ate 3 roaches and 5 spiders. Doris saw Bill eating them, so she ate 4 roaches and 7 spiders.

   How many roaches did the children eat altogether?____

   How many roaches were left in the box?____

   How many spiders were left in the box?____

   How many total animals were left in the box?____

6 ADDITION & SUBTRACTION

9. In the kindergarten classroom, a small round tank had 3 snakes, a large round tank had 8 snakes, a small square tank had 4 snakes, and a large square tank had 10 snakes.

    Draw all the tanks and show the snakes in each.
    All the snakes in the square tanks got away.

    How many snakes got away?____

    How many snakes are left in the tanks?____

10. Mario earned $95 mowing lawns. He spent $30 for a fishing pole. Then he found $15 in a fish he caught.

    Now how much money does he have?____

Lesson 4

1. Brenda put 3 turtles and 4 frogs in her brother's bed.

    How many total animals were in the bed?____

2. Monster had 11 cow ears. He ate 5 ears for dinner.

    How many were left?____

3. The Dolphins football team had 13 points. Then they scored a touchdown which gave them 6 more points.

    How many total points did they have?____

4. There are 26 apples in a bag. Eight apples have worms in them.

    How many apples do not have worms?____

5. Mr. Davis had 106 hairs on his head. One morning he found that 30 new hairs had grown on his head. Then he became a teacher and lost 80 hairs worrying about his students.

    How many hairs did he have left?____

6. Alfred had 45 hotdogs in his refrigerator. He ate 18 hotdogs and his sister ate 22 hotdogs.

   How many hotdogs were left?____

7. 45 people were planning to go on a picnic. Then 3 men and 4 women decided not to go. The rest of the people did go on the picnic.

   How many people went on the picnic?____

8. Mrs. McRae is a teacher who had 35 dog noses in a jar. For Valentine's Day, Ricardo gave her 14 more noses. Then Mrs. McRae gave 2 noses to each of the 5 best children in her class.

   How many noses did Mrs. McRae have left?____

9. Shana weighed 120 pounds. During the Christmas season she gained 65 pounds. Then she went on a diet and dropped 30 pounds. But in March she started eating again until she gained 90 pounds.

   After that, how much did she weigh?____

Lesson 5

1. How many more forks than spoons are in the picture?____

   5 forks
   −3 spoons
   2 more forks than spoons

2. How many more fish than frogs are there?____

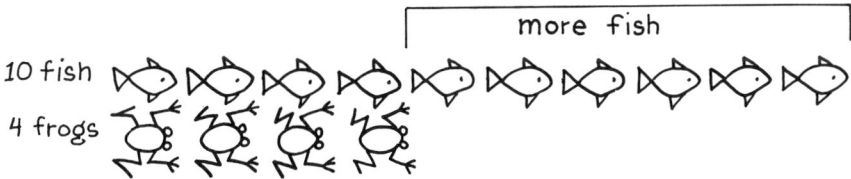

8   ADDITION & SUBTRACTION

3. Bob had 8 marbles.

    Write Bob's name and draw 8 marbles in a row.

    Dan has 5 marbles.

    Write Dan's name under Bob's name.

    Draw Dan's 5 marbles under Bob's marbles.

    How many more marbles does Bob have than Dan?____

4. A room had 8 chairs. There are 6 people in the room.

    How many more chairs than people are there?____

5. There are 7 sodas and 11 people at a picnic.

    How many people cannot have soda?____

6. A small box has 4 cookies and a big box has 16 cookies.

    Which box would you prefer?____

    Why do you prefer that box?____

    How many more cookies does the big box have than the small box?____

7. Mr. Gorilla and Mr. Frog are learning to talk. Mr. Gorilla knows 18 words. Mr. Frog knows 11.

    How many more words does Mr. Gorilla know than Mr. Frog?____

8. Andy can go home the short way that takes 10 minutes. But he has to pass the body snatcher's house and he might get him. The safe way home takes 23 minutes.

    How many more minutes does the safe way take?____

9. A small box has 6 pencils. A large box has 10 pencils.

   How many pencils are there in two small boxes?____

   How many more pencils are there in two small boxes than in one large box?____

10. A small box has 6 cookies. A large box has 15 cookies.

    How many more cookies are there in 3 small boxes than 1 large box?____

11. There are 20 pages in a coloring book. Maria colored 6 pages and Phil colored 10 pages.

    How many total pages did the children color?____

    How many pages in the book were not colored?____

    How many more pages did Phil color than Maria?____

12. Cathy had 17 dolls. Sue only had 3 dolls. Cathy gave 5 dolls to Sue.

    How many dolls does Cathy have now?____

    How many dolls does Sue have now?____

    How many more dolls does Cathy have than Sue?____

13. Paul had 20 music tapes and John had 10 tapes. Paul gave 1 tape to John.

    Now how many more tapes did Paul have than John?____

14. M&M's are sold in small bags and large bags.

    If you were very hungry, which size would you prefer to be given?____ Why?____
    Would you prefer to get 2 small bags or 1 large bag?
    What other information do you need before deciding?
    Write your answer on your paper.

10 ADDITION & SUBTRACTION

Lesson 6

1. Lucy is 42 inches tall. Cindy is 7 inches taller.

   How tall is Cindy?____

2. The fat lady in the circus weighs 410 pounds more than Roger. Roger weighs 81 pounds.

   How much does the fat lady weigh?____

3. Andy has 48 comic books. Kevin has 32 comic books.

   Which boy has more comic books?____

   How many more comic books does Andy have than Kevin?____

4. Bobby has 18 worms. Judy has 4 worms and Sally has 7 worms.

   How many more worms does Bobby have than Judy and Sally together?____

5. Dan has 11 cats and 15 dogs. Chris has 2 rabbits, 6 ducks, and 8 goldfish.

   How many more pets does Dan have than Chris?____

6. Marbles are sold in boxes containing 15 marbles. They are also sold in cans containing 25 marbles.

    Would you get more marbles in 3 boxes or 2 cans?____

    How many more marbles would you get in 2 cans than in 3 boxes?____

7. Paul had 30 mice and John had 8 mice. Paul gave 10 of his mice to John.

    How many mice does Paul now have?____

    How many mice does John now have?____

    How many more mice does Paul have than John?____

8. James had 205 baseball cards and Pete had 150 cards. James gave Pete 10 cards.

    Now how many more cards does James have than Pete?____

9. Steven had 16 bees. He gave 5 bees to Bill. Bill ate 3 bees in a salad.

    Now how many more bees does Steven have than Bill?____

Lesson 7

1. Joe runs 4 miles a day. Lucy runs 3 miles a day more than Joe.

    How many miles does Lucy run in one day?____

    How many miles does Lucy run in three days?____

2. Kelly watches 3 hours of TV a day. Betty watches 2 more hours a day than Kelly.

    For how many hours does Betty watch TV in two days?____

3. There were 35 ants in a can. Carlos ate 8 ants and Marcia ate 17 ants.

    How many total ants did the children eat?____

    How many ants were left in the can?____

    How many more ants did Marcia eat than Carlos?____

12 ADDITION & SUBTRACTION

4. There are 53 seats in a room where a movie was shown. One day 19 men, 12 women, and 7 children went to the movie.

   How many adults went to the movie?____

   How many more adults than children went to the movie?____

   How many total people went to the movie?____

   How many seats in the room were empty?____

5. Rick and Tiffany have to eat 18 toads to win a contest. In the morning Rick ate 3 toads and Tiffany ate 8.

   How many do they still need to eat?____

6. Keith earned $85. He gave $20 to his parents. He put the rest in the bank. He already had $110 in the bank.

   How much does he have in the bank now?____

7. On a trip Dwayne spent 4 days in Chicago, 7 days in New Orleans, and 5 days in Miami. It rained for 2 days in Chicago, 1 day in New Orleans, and 2 days in Miami.

   How many dry days did he have in the 3 cities altogether?____

8. Ted always ate 5 more hotdogs than his sister. One day she decided to trick him and ate 7. When Ted got to 10 he exploded.

   How many more did he have left to eat?____

9. Bobby wants to buy a used bike for $25. He only has $13.

   How much more money does he need?____

10. Darlene has $10. She wants to buy a doll for $6 and a baseball bat for $12.

    How much more money does she need?____

Lesson 8

1. Paul opened a box of candy. His dog quickly ate one candy. But 9 candies were left.

   How many candies were in the box before the dog ate one?____

2. Jack caught some spiders. His cat ate 2. Then Jack had 4 left.

   How many had Jack caught?____

3. Gillian bought a cookie costing 7¢. She gave the clerk a coin, and the clerk gave Gillian 3¢ change.

   What coin did Gillian give the clerk?____

4. Chase found some cash in an old sofa. He gave $3 to his little sister. Then he had $8 left.

   How much had he found?____

5. Mrs. Booklover bought a box of old books. Pru Booklover ate 3, and Gru Booklover ate 7. Then 20 books were left.

   How many books were in the box before the Booklovers ate some?____

6. Mr. Sizzz bought an elephant for $105 to guard his house. He gave the circus owner some money and got $15 change.

   How much money had he given the owner?____

7. Queen Loku had too many servants, and they were driving her crazy. She gave 8 to her cousin and 12 to her sister. Then she had 4 good ones left.

   How many did she have before giving some away?____

8. Jeff took 3 lucky shark teeth from a box and Alvin took 5. Then there were 13 left.

   How many shark teeth were in the box before the boys took some?____

9. A shirt cost $5 and a pair of slacks cost $8. Alex got $7 change after he bought the shirt and slacks.

   How much money did he give the clerk?____

ADDITION & SUBTRACTION

Lesson 9

Delicious Dining

worms
10¢ each

green fried eggs
25¢ each

spiders
60¢ each

potato soda
30¢ a can

wolf-spit tea
15¢ a cup

snake salad
$2.50

roast rat tails
35¢ each

skunk ears
40¢ each

1. Bill bought 2 worms and a spider. He gave the clerk $1.00.

   How much change did he receive?____

2. Paul bought one skunk ear. Lisa bought three cans of potato soda.

   How much more did Lisa spend than Paul?____

3. Alison bought a skunk ear and something else, but I could not see what. The clerk charged Sue 65¢.

   What was the other thing Alison bought?____

4. Keith bought 2 rat tails and one other thing. The clerk charged Keith $1.30.

   What was the other thing Keith bought?____

5. Judy wants to eat a skunk ear and a snake salad. She has $1.30.

   How much more money does she need?____

6. Pogo bought 2 spiders and a skunk ear. He gave the clerk all the money in his pocket and got $4.10 change.

   How much money had he given the clerk?____

7. Here is the recipe for Stinky Stew:
   Stinky Stew
   10 worms
   2 roast rat tails
   5 skunk ears
   2 cups of wolf-spit tea

   If you bought these ingredients and gave the clerk $10, how much change would you get?____

ADDITION & SUBTRACTION 15

8. Make up a recipe from the above menu for enough super soup to feed seven teachers at a picnic.

   Write the recipe on your paper with this heading:
   Super Soup Recipe

   How much will your super soup cost?____

Bonus Question: Make up a list of food with prices like that above. Then create a recipe and let another student find out how much it would cost.

Lesson 10

1. The distance from Tom's house to Bob's house is 8 miles. The distance from Tom's house to school is 13 miles.

   Draw this picture on your paper. Include the numbers and question mark.

   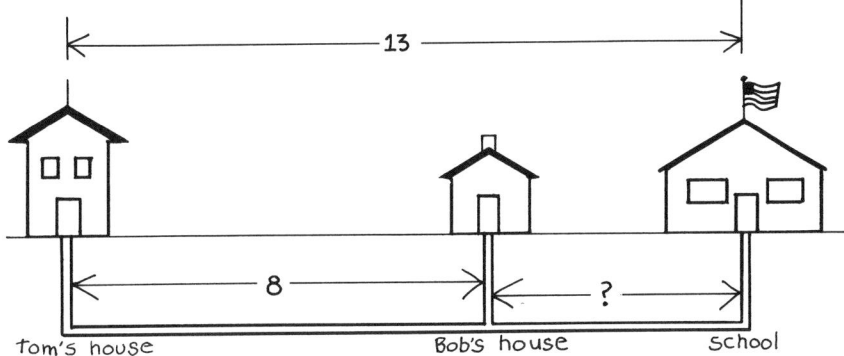

   What is the distance from Bob's house to school?

   Write your answer by the question mark on your picture.

2. The distance from Pete's house to Monster's house is 3 miles. The distance from Pete's house to Judy's house is 12 miles.

   Draw this picture on your paper.

   Write the distances by the question marks.

   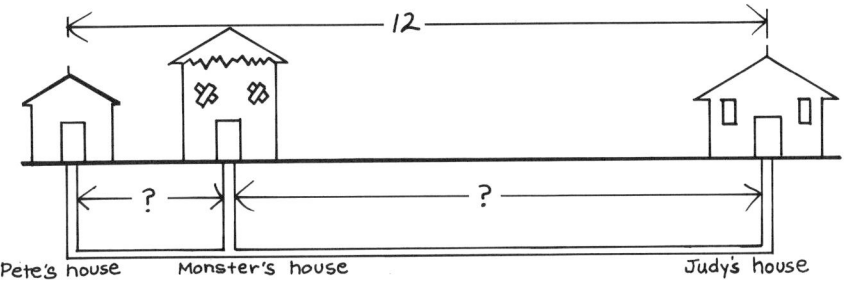

   Who lives farther away from Monster?____

16 ADDITION & SUBTRACTION

3. Patty is 5 miles from home. The lion is 4 miles from his cave. Patty's house is 11 miles from the lion's cave.

   Draw this picture and write the distances by the question marks.

4. Starting at one corner, the borders of a park run 5 miles north, 6 miles east, 5 miles south, and back to the starting corner.

   Draw this picture of the park and write the distances by the question marks.

   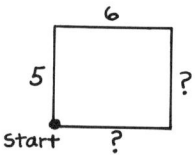

   What is the total distance around the park?____

5. Starting at one corner, the borders of a park run 10 miles west, 4 miles north, 3 miles east, 5 miles north, 7 miles east, and back to the starting corner.

   Draw this picture of the park and write the distances by the question marks.

   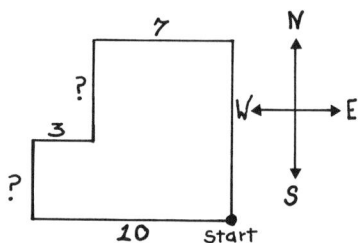

   What is the total distance around the park?____

6. Starting at one corner, the borders of a park run 3 miles north, 7 miles west, 3 miles south, and back to the starting corner.

   Draw the park with the lengths written on each side.

ADDITION & SUBTRACTION   17

7. Starting at one corner, the borders of a park run 3 miles north, 2 miles west, 3 miles north, 5 miles west, 6 miles south, and back to the starting corner.

   Draw the park with the length written on each side.

   What is the total distance around the park?____

Lesson 11

1. Ryan paid $15 for a toy boat and a truck. The boat cost $9.

   Copy this picture and write the price of the truck by the question mark.

   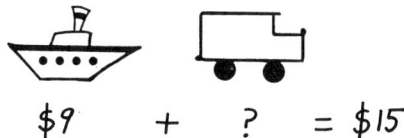

2. Beverly paid $11 for 2 ribbons and a hat. Each ribbon cost $3.

   Copy this picture and write the price of the hat by the question mark.

3. My dog Spot jumped out of our car 70 miles from our house. It took him 3 days to run home. The first day he ran 30 miles and the second day 26 miles.

   Copy this picture and write the distance he went the third day by the question mark.

   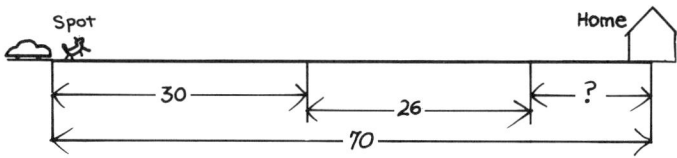

4. Kenny paid 70¢ for 3 crocodile teeth and a skunk foot. Each tooth cost 15¢.

   How much did the skunk foot cost?____

5. A theatre has 400 seats. Last night 120 adults and 170 children went to see a horror movie. It was so frightening that 23 adults and 11 children left.

   How many seats in the theatre were empty?____

18   ADDITION & SUBTRACTION

6. A car dealer had 32 Fords and 18 Hondas on his lot. One night 6 Fords and 11 Hondas were stolen. The radios were torn out of 5 other cars.

   How many cars remained on the lot undamaged?____

7. Mrs. Clean put 20 pieces of laundry out to dry on the clothesline. A goat ate 4 shirts, 3 socks, 2 girdles, and a skirt. But one of the girdles tickled its stomach and made it throw up 2 shirts and 1 girdle undamaged.

   How many pieces of laundry did Mrs. Clean lose?____

8. Tom and Fred had to deliver flowers to 33 homes on Thursday. In the morning Tom delivered to 8 homes and Fred delivered to 9 homes.

   How many homes must they deliver flowers to on Thursday afternoon?____

Lesson 12

1. Tickets for a movie cost $6 each. Ted wants to take his sister, but he only has $8.

   How much must his sister bring so they both can go?____

2. Mrs. Dora Smith spent $46 for a tire and $13 for gas. Her son spent $53 for repairs. Mrs. Smith gave the attendant $95.

   How much must her son give the attendant to pay the rest of the total bill?____

3. Dan is a paperboy with 93 customers. Twenty customers get papers on weekends only, and 15 get papers on weekdays only.

   How many customers get papers on all 7 days of the week?____

4. One policeman shot 5 bullets at a robber's car and another policeman shot 3 bullets.
   Which line shows how many total bullets the two policemen shot?

   a. 5−3=____

   b. 5+3=____

   c. 5×3=____

   d. 3−5=____

ADDITION & SUBTRACTION 19

5. Avis made 42 frog pies. She gave her aunt 8 pies.

   Which line shows how many pies Avis had left?

   a. 8+42=____

   b. ____+42=8

   c. 42−8=____

   d. 8×34=____

6. Jake and Phil have 58 marbles together. Jake owns 23 of the marbles.

   How many does Phil own?____

7. Judy and her sister Phylis have 60 dresses together. Judy owns 25 of the dresses.

   Which line shows how many dresses Phylis owns?

   a. 25+60=____

   b. 60−25=____

   c. ____+60=25

8. Christine found 5 skirts she liked at the store. Their prices were $6, $11, $14, $8, and $7. She bought the 3 cheapest skirts.

   How much did she pay altogether?____

9. A pet store has 17 dogs; 7 dogs are black, 4 dogs are white, and 6 dogs are brown.

   How many dogs are not brown?____

Lesson 13

1. Cathy has 18 dolls. Paula has 5 fewer dolls than Cathy.

   Copy this picture.

   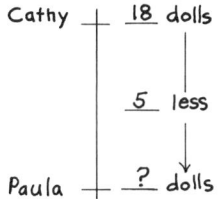

   Write the number of dolls Paula has by her name on your picture.

20 ADDITION & SUBTRACTION

2. Andy has $25. Beth has $12 more than Andy.

   Copy this picture.

   Write the amount of money Beth has next to her name on your picture.

3. Ed weighs 80 pounds. This is 10 pounds more than he weighed last year.

   Copy this picture.

   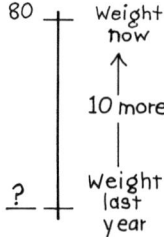

   Write how much Ed weighed last year by the question mark on your picture.

4. Bill has been on a diet. He weighs 92 pounds. This is 4 pounds less than he weighed last month.

   Copy this picture.

   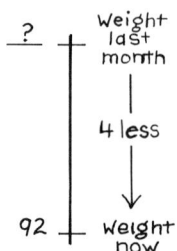

   Write how much Bill weighed last month by the question mark on your picture.

5. Pat owns 7 toy trucks. He owns 3 more trucks than Larry.

   Copy this picture.

   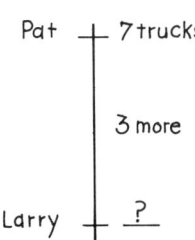

   Write the number of trucks Larry owns by his name on your picture.

6. Football Fred owns 8 cars. He owns 3 more cars than Tennis Tom.

   Copy this picture.

   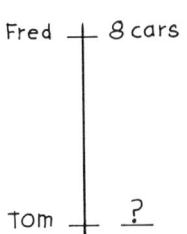

   Write the number of cars Tom owns by his name on your picture.

7. The temperature in Miami is 80. This is 12 degrees warmer than the temperature in Atlanta.

   Copy this picture.

   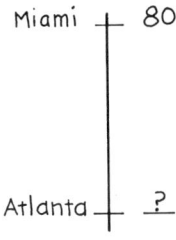

   Write the temperature of Atlanta by its name on the diagram.

8. If Mark had $2 more he would have $10.

   How much does Mark have now?____

22 ADDITION & SUBTRACTION

9. If Scott had 18 more ants, he would have 100 ants.

   How many ants does he have now?____

10. On Halloween, Sally saw 5 ghosts and Joey saw 7 ghosts. Paul saw 8 more ghosts than Sally and Joey combined.

    How many ghosts did Paul see?____

Lesson 14

1. Starting at one corner, the borders of a park run 8 miles south, 9 miles west, 8 miles north, and back to the starting corner.

   Draw a picture of the park with the length written on each side.

   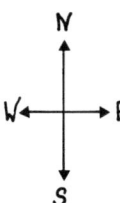

   Wolfman chased Pam all the way around the park.

   How far did he chase her?____

2. Starting at one corner, the borders of a park run 6 miles south, 3 miles west, 5 miles south, 7 miles west, 11 miles north, and back to the starting corner.

   Draw this picture of the park. The lengths are written on 2 sides.

   Write the lengths on all the sides on your picture.

   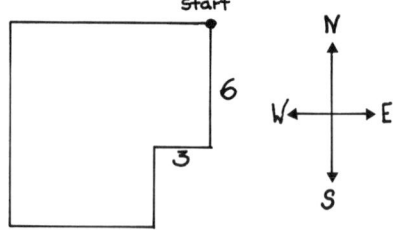

   If you run completely around the park, how far would you run?____

ADDITION & SUBTRACTION   23

3. Starting at one corner, the borders of a park run 14 miles north, 12 miles west, 7 miles south, 8 miles east, 7 miles south, and back to the starting corner.

   Draw a picture of the park with the lengths written on each side.

   What is the total distance around the park?____

4. Starting at one corner, the borders of a park run 3 miles north, 4 miles east, 1 mile south, 5 miles east, 1 mile north, 4 miles east, 3 miles south, and back to the starting corner.

   Draw a picture of the park with the length written on each side.

   What is the total distance around the park?____

5. The total distance around any figure is called the PERIMETER.

   What is the perimeter of this figure?____

   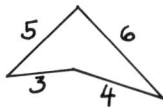

   What is the perimeter of this figure?____

   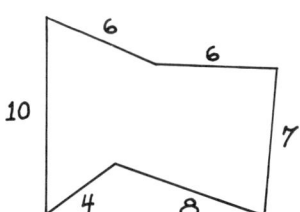

Lesson 15

1. Jack has $125. He wants to buy 2 pet alligators which cost $70 each.

   How much more money does he need?____

2. A store has 160 customers. They bought 125 ham sandwiches, 135 cheese sandwiches, and 115 tuna sandwiches.

   How many total sandwiches did they buy?____

24 ADDITION & SUBTRACTION

3. The school library has 937 books. The fourth grade students checked out 129 books.

   How many books were left?____

4. Gayle bought a pair of stockings costing $3.49. She gave the clerk $5.

   How much change did she receive?____

5. Christine bought the following snacks for her family at the circus: 3 slices of pizza, 4 orange drinks, and 2 candy apples. Pizza is $2 a slice, orange drinks are 75¢ each, and candy apples are $1.50 each.

   How much change did Christine get from a $20 bill?____

6. Two weeks ago, Al saw a bike for $187. When he went back to buy it yesterday, it was on sale for $148.

   How much money did he save?____

1 Additional Exercises

Lesson 1

1. Calvin had 5 devil fish.

   Draw the 5 fish. Here is one.

   His cat ate 2 fish.

   Cross out 2 fish on your paper.

   How many fish did Calvin have left?____

   Do this subtraction on your paper.   5
                                        −2

2. Denise had 4 turtles.

   Draw 4 turtles. Here is one.

   She bought 3 more turtles.

   Draw 3 more turtles on your paper.

   How many total turtles does Denise have?____

   Do this addition on your paper.   4
                                     +3

25

26 ADDITIONAL EXERCISES

3. Lisa had 4 friends.

Draw 4 children. Here is one.

One boy moved to Geneva.

Cross out 1 boy on your paper.
Then Lisa got 2 new friends.

Draw 2 new children.

Now how many friends does Lisa have?____

Do these problems on your paper.   4   3
                                                                                                                      −1   +2

4. Paul had 7 frogs.

He gave 3 frogs to his friend Tom.

How many did Paul have left?____
Then Paul caught 1 frog.

Now how many frogs does Paul have?____

Do these problems on your paper.   7   4
                                                                                                                       −3   +1

5. Lilly had 10 spiders.

Two spiders got away.

How many spiders were left?____
Then 3 new spiders crawled into Lilly's window to live with her.

Now how many spiders did Lilly have now?____

6. Bobby had $4. $ $ $ $

   On Monday Bobby's father gave him $2.

   Then how much did Bobby have?____
   On Tuesday Bobby spent $1 for a toy.

   How much money did he have left?____

7. A train had 9 passengers.

   At one stop 3 passengers got off.

   How many passengers were left on the train?____
   At the next stop 2 people got on.

   Now how many people were on the train?____

8. Ken had 12 ants. He ate 3 ants. Then he caught 5 ants.

   How many ants did he have?____

9. Jason has $15. He spent $4 on perfume for his girlfriend. Then he earned $9 washing cars.

   Now how much money does he have?____

28 ADDITIONAL EXERCISES

Lesson 2

1. Here is a box with 9 candies. John ate 3 candies. Cover 3 candies with your finger.

   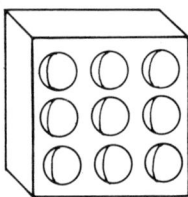

   How many candies were left in the box?____

   Do this subtraction.   9
                                     −3

2. Paula had 5 pencils. Then she bought 2 more pencils.

   Hold 2 fingers next to the pencils in the picture.

   How many total pencils did Paula have?____

   Do this addition.   5
                             +2

3. You won $6 in a contest.

   Draw $6 on your paper.

   You spent $2 for a hot fudge sundae.

   Cross out $2 on your paper.

   How much did you have left?____
   Then you earned $3 selling magazines.

   Draw $3 more on your paper.

   Now how much money do you have?____

4. Jason had 15 marbles.

   Draw 15 marbles on your paper.

   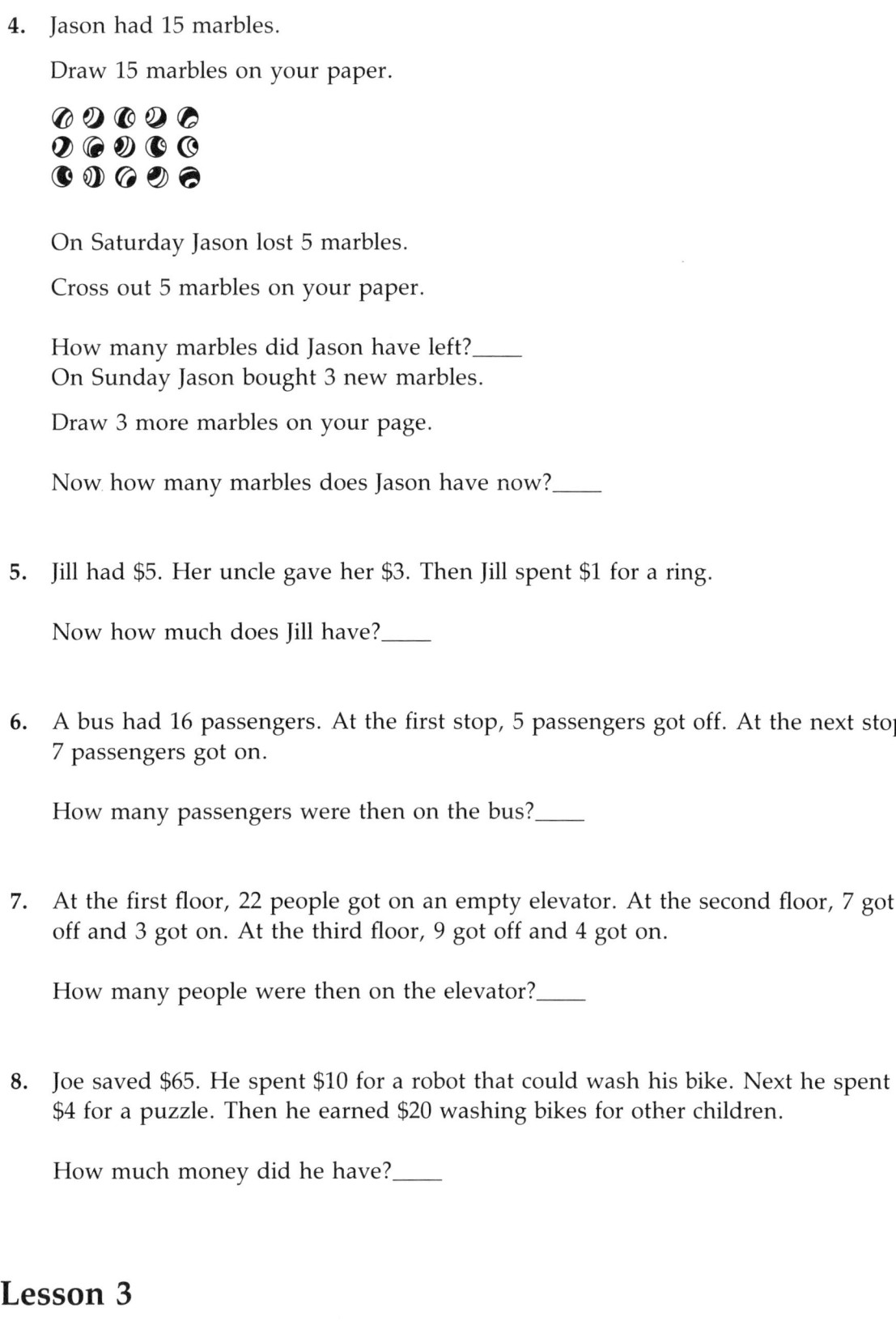

   On Saturday Jason lost 5 marbles.

   Cross out 5 marbles on your paper.

   How many marbles did Jason have left?____
   On Sunday Jason bought 3 new marbles.

   Draw 3 more marbles on your page.

   Now how many marbles does Jason have now?____

5. Jill had $5. Her uncle gave her $3. Then Jill spent $1 for a ring.

   Now how much does Jill have?____

6. A bus had 16 passengers. At the first stop, 5 passengers got off. At the next stop, 7 passengers got on.

   How many passengers were then on the bus?____

7. At the first floor, 22 people got on an empty elevator. At the second floor, 7 got off and 3 got on. At the third floor, 9 got off and 4 got on.

   How many people were then on the elevator?____

8. Joe saved $65. He spent $10 for a robot that could wash his bike. Next he spent $4 for a puzzle. Then he earned $20 washing bikes for other children.

   How much money did he have?____

Lesson 3

1. A cat weighed 14 pounds before lunch. It ate 2 pounds of cat food for lunch.

   How much did it weigh after lunch?____

30 ADDITIONAL EXERCISES

2. A snake weighed 20 pounds. It swallowed a 3-pound rat.

   Then how much did the snake weigh?____

3. Marie had 8 dolls.

   Her brother threw 3 dolls in the garbage.

   How many dolls did Marie have left?____

4. A bad alligator that chased children weighed 26 pounds. A dog bit off 4 pounds of the alligator's tail.

   Then how much did the alligator weigh?____

5. Charlie had 17 baseball cards. He lost 6 cards while he was out playing.

   How many cards did he have left?____

6. A cartoon Monster had 15 fingers in a jar. He ate 5 fingers. Then Madman gave him 7 fingers.

   How many fingers did Monster have?____

   Do these problems on your paper.   15   10
                                      −5   +7

7. Judy had 18 cookies. Her mother gave her 6 more cookies. But a mouse ate 3 cookies.

   Then how many cookies did Judy have?____

8. Paul had 21 slimeballs. Jack gave him 6 more. But Paul lost 3 at the park.

   Then how many slimeballs did Paul have?____

   Do these problems on your paper.   21   27
                                      +6   −3

9. Debbie had 27 frisbees. She threw 4 at Carla, and she threw 9 at Neal.

   Now many frisbees did Debbie have left?____

10. Monster had 18 skunk ears in a jar. For his birthday, Madman gave him 6 ears, and Beast gave him 11 ears.

    Then how many ears did Monster have?____

Lesson 4

1. Dr. Frite had 8 rocks in a tank.

   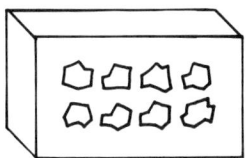

   He used 3 rocks to make monsters.

   How many rocks were left in the jar?____

2. One night Loren ate 6 garbage cans, David ate 7 garbage cans, and Cedric ate 9 cans.

   How many garbage cans did they eat altogether?____

3. There were 8 frogs in a pond.

   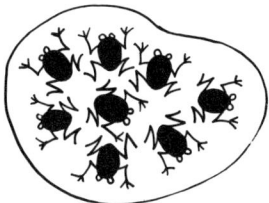

   Bill caught 2 of the frogs. Judy caught the rest.

   How many did Judy catch?____

4. Linda had 14 turtles. Her father gave 3 to a friend. Her pet took 5 others. But Linda bought 2 new turtles.

   Then how many turtles did Linda have?____

32   ADDITIONAL EXERCISES

5. Isaac had 2 cages with 4 skunks in each cage. His mother cooked 3 skunks for dinner.

   Then how many skunks did Isaac have left?____

6. Shana had 14 jars of worm guts. She gave 4 jars to Pee Wee. But her friend Dawn gave her 8 new jars.

   Then how many jars did Shana have?____

7. Cedric had 8 cars. Mike stole 3 cars. But Cedric got 6 new cars.

   Now how many cars does Cedric have?____

8. Mrs. Doyle baked 40 cookies. Her son ate 7 cookies, and her daughter ate 4.

   How many cookies were left?____

9. Ben had 12 Hot Wheels. Three Hot Wheels burned when left near the fire. And 4 Hot Wheels were taken when his car was robbed. But Ben got 8 new Hot Wheels for his birthday.

   Then how many Hot Wheels did Ben have?____

Lesson 5

1. Andrea had 8 cats. Her mother found other homes for 2 cats. But Janet gave Karen 1 new cat.

   Then how many cats did Karen have?____

2. There were 6 roaches and 2 spiders in Carl's bed. Then 2 roaches and 1 spider jumped out of the bed.

   How many animals were left in the bed?____

3. Bob has 5 race cars. Tom has 2 race cars.

   How many cars do Bob and Tom have altogether?____

   How many cars should Tom buy to have the same number as Bob?____

4. Joe bought a toy boat for $3 and a car for $1.

   How much did he spend altogether?____

5. Edith spent $4 to buy a doll and a ribbon. The doll cost $3.

   How much did the ribbon cost?____

   Do this subtraction on your paper.   4
   $$\underline{-3}$$

6. James bought a toy soldier and a ball. The clerk charged him $7 total. The ball was $2.

   How much was the soldier?____

   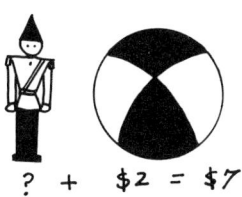

   ? + $2 = $7

7. Janice found 7 seashells at the beach. Jim found 3 shells.

   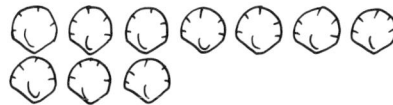

   How many more shells must Jim find to have the same number as Janice?____

   Do this subtraction on your paper.   7
   $$\underline{-3}$$

34 ADDITIONAL EXERCISES

8. A mailman must deliver 15 packages today. He delivered 10 packages before lunch.

   How many packages must he deliver after lunch?____

9. Eight children were in the pool. Three more children got in. Then 5 children got out.

   How many children were left in the pool?____

10. Terry spent $5 for a light for his bike and $2 for a horn. He gave the clerk $10.

    How much change did he receive?____

Lesson 6

1. A room has 5 chairs. Seven people came into the room.

   How many more chairs are needed for all the people to be able to sit?____

2. A room has 10 chairs. Only 3 people came into the room.

   How many extra chairs are there?____

3. A small box has 3 pencils and a big box has 5 pencils.

   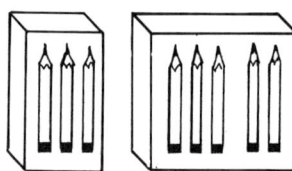

   How many more pencils are in the big box than in the small box?____

4. Paul wants to buy a toy for $6. He only has $4.

   How much more money does he need?____

5. Keith has $7. Bob has $4.

   How much more money does Keith have than Bob?____

6. A small box has 6 candies. A big box has 9 candies.

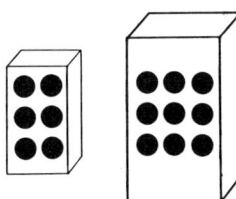

How many more candies are there in the large box than in the small box?____

7. Rosa has 5 gold stars. Ann has 2 gold stars.

How many more stars does Rosa have than Ann?____

8. Jackie ate 5 apples. Wilner ate 1 apple.

How many more apples did Jackie eat than Wilner?____

9. Tim drank 3 bottles of bug juice. Joe drank 2 more bottles than Tim.

How many bottles did Joe drink?____

Do this addition on your paper.   3
                                                +2

10. Marie has 6 dolls.

Draw Marie's dolls on your paper. Here is one doll.

Nancy has 3 more dolls than Marie.

Draw all of Nancy's dolls.

How many dolls does Nancy have?____

ADDITIONAL EXERCISES

11. Alex weighs 82 pounds. Hayes weighs 70 pounds.

    How much must Hayes gain to weigh as much as Alex?____

12. A giant snake is 28 inches long. A fish is 5 inches long.

    How much bigger is the snake?____

13. An alligator weighs 32 pounds. My dog weighs 22.

    How much more does the alligator weigh than my dog?____

2 Multiplication

Lesson 1

1. Cindy has 3 jars. Each jar has 2 butterflies.

   Draw 3 jars and draw 2 butterflies in each jar. Here is one jar.

   What numbers can you multiply to find how many butterflies Cindy has?

   Write the multiplication on your paper.

2. Kevin has 4 tanks. Each tank has 3 fish.

   Draw the 4 tanks and draw 3 fish in each tank.

   What numbers can you multiply to find how many fish Kevin has?

   Write the multiplication on your paper.

3. Each box has 2 cupcakes. There are 5 boxes.

   Draw the five boxes and draw 2 cupcakes in each box.

   What numbers can you multiply to find the total number of cupcakes?

   Write the multiplication on your paper.

38 MULTIPLICATION

4. Each coffin has 4 spiders. Dracula has 6 coffins.

   Draw the 6 coffins and write the number 4 in each. Here is one coffin.

   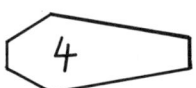

   How many total spiders does Dracula have?____

5. Nina has 4 bags. Each bag has 11 lollipops.

   Draw the 4 bags and write 11 in each bag.

   How many lollipops does Nina have?____

6. Ozzie wanted a full education. He ate 4 books. Each book had 100 pages.

   Draw the 4 books and write 100 in each. Here is one book.

   How many pages did Ozzie eat?____

7. Each magic box has 25 crocodile teeth. There are 4 magic boxes.

   Draw the 4 boxes and write 25 in each.

   What is the total number of teeth in all the magic boxes?____

8. Each cage has 25 white mice and there are 5 cages.

   How many white mice are there altogether?____

9. Each bus holds 40 children. There are 9 full buses.

   How many children are there altogether?____

Lesson 2

1. Mrs. Lark bought 5 boxes of chocolate-covered grasshoppers for her party. Each box has 10 grasshoppers.

   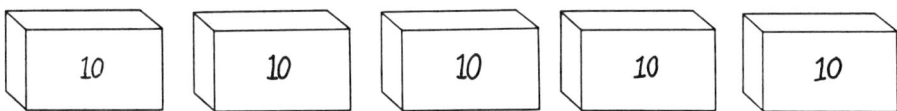

   How many total grasshoppers does Mrs. Lark have for her party?____

2. Mr. Lark bought a big box of grasshoppers and a small box. The big box has 10 grasshoppers and the small box has 5.

   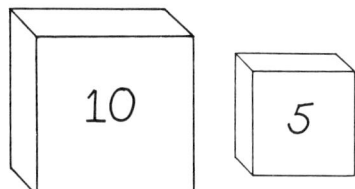

   How many total grasshoppers did Mr. Lark buy?____

3. Mr. Bumphead bought one big can of dried lizards and one small can. The big can has 15 lizards and the small can has 7.

   Draw the cans and write the number of lizards in each.

   What is the total number of lizards Mr. B. bought?____

4. Mrs. Bumphead bought 7 cans of lizards. Each can had 15 lizards.

   Draw all the cans and write the number of lizards in each.

   How many total lizards did Mrs. B. buy?____

5. Jesse bought 4 boxes of pencils. Each box had 30 pencils.

   Draw all the boxes and write the number of pencils in each.

   How many total pencils did Jesse buy?____

40 MULTIPLICATION

6. Tanya bought one large box of pencils and one small box. A large box has 30 pencils and a small box has 4 pencils.

   Draw the boxes and write the number of pencils in each.

   How many total pencils did Tanya buy?____

7. A monster trapped 20 children a day. He did this for 5 days.

   How many total children did he trap?____

8. Guy Good trapped 20 monsters one day. The next day he trapped 5 monsters.

   How many total monsters did he trap?____

9. Max Mean crushed 3 big boxes of baby chicks and one small box. Big boxes contain 12 chicks and small boxes contain 8.

   Draw all the boxes and write the number of chicks in each.

   How many total chicks did Max crush?____

Lesson 3

1. Billy broke 5 jars. Each jar contained 30 jelly beans.

   Draw the jars and write the number of beans in each one.

   How many jelly beans were in the jars altogether?____

2. Clint bought 3 radios. Each radio needs 4 batteries. He also bought 2 toy trucks. Each truck needs 5 batteries.

   Draw all the radios and trucks and write the number of batteries in each. Here is a radio and a truck.

   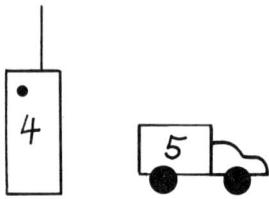

   How many batteries are needed for all the radios and trucks?____

3. A teacher bought 3 toy boats and 5 talking dolls for his students. Each boat requires 2 batteries and each doll requires 4.

   Draw all the boats and dolls, and write the number of batteries in each.

   How many batteries must the teacher buy?____

4. In the pet store, there are 2 round tanks and 3 square tanks. Each round tank has 10 fish and each square tank has 8 fish.

   Draw all the tanks and write the number of fish in each tank.

   How many fish are there altogether?____

5. Janice had 2 jars with 11 spiders in each jar. Her cat ate 14 spiders.

   How many spiders did Janice have left?____

6. Kerri had 4 friends. Each friend gave her 5 jelly beans to eat for lunch. She kept 7 of the beans and ate the rest.

   How many beans did she eat?____

7. Mario had 5 boxes with 4 grasshoppers in each box. He put 3 grasshoppers in the tub with his brother. He also put 8 grasshoppers in his sister's bed.

   How many grasshoppers did he have left?____

8. At camp there are 10 cabins. Each cabin has 15 children. One night bears grabbed 8 children. Also, witches flew away with 12 children. But in the morning 7 new children came to the camp.

   Then how many children were at the camp?____

Lesson 4

1. Each round box has 8 candies and each square box has 12 candies. There are 3 round boxes and 5 square boxes.

   Draw all the round boxes and square boxes and write the number of candies in each box.

   How many candies are there in the round boxes?____

   How many candies are there in the square boxes?____

   How many candies are there in all the boxes?____

42 MULTIPLICATION

2. Mrs. Smith bought 4 talking teddy bears and 6 portable TVs. Each bear requires 2 batteries and each TV requires 4 batteries.

   Draw the 4 bears and 6 TVs and write the number of batteries in each one.

   How many batteries does Mrs. Smith need for all the bears and TVs?____

3. Janet bought 8 large boxes of pens. Each large box had 50 pens. She also bought 5 small boxes of pens. Each small box had 30 pens.

   How many pens did she buy?____

4. Thelma wants to make 2 bracelets. Each bracelet requires 15 beads. She also wants to make 4 necklaces. Each necklace required 32 beads.

   How many beads does she need altogether?____

5. Each man eats 3 hotdogs, each woman eats 2 hotdogs, and each child eats 1 hotdog.

   How many hotdogs are needed for a picnic attended by 10 men, 6 women, and 20 children?____

6. Mrs. Jones bought 11 flashlights, 10 talking clowns, and 5 race cars for the children in her class. Each flashlight requires 2 batteries, each clown requires 3 batteries, and each car requires 4 batteries.

   How many batteries should Mrs. Jones buy?____

7. Rodney wants to bake 4 cookies and 3 cakes. Each cookie requires 2 cherries and each cake requires 10 cherries.

   How many cherries does he need?____

8. Each lion eats 25 pounds of meat, each leopard eats 15 pounds, and each dog eats 2 pounds.

   How much meat is needed to feed 6 lions, 9 leopards, and 12 dogs?____

Lesson 5

1. Five boys each ate 3 ants.

   How many ants did they eat altogether?____

2. Eleven boys each ate 3 ants.

   How many ants did they eat altogether?____

3. Twenty giant ants each ate 3 boys.

   How many boys did they eat altogether?____

4. Eight children each have 2 cats.

   How many cats do they have altogether?____

5. Twenty termites moved into my house. Each termite ate 4 wooden boards.

   How many boards did they eat altogether?____

6. Fifteen goats each ate 2 tires at a car lot one night.

   How many tires did they eat altogether?____

7. Six girls and 4 boys each ate 3 doughnuts.

   How many doughnuts did they eat altogether?____
   READ THIS: The words "twice as many" mean 2 times as many.

8. Nine girls and twice as many boys went to the zoo.

   How many boys went to the zoo?____

44 MULTIPLICATION

9. Six mice and twice as many rats live in my cellar.

   How many rats live in the cellar?____

10. Four lions and twice as many tigers escaped from the circus.

    How many animals escaped altogether?____

Lesson 6

1. Each robot needs 3 batteries and each car needs 5 batteries. There are 6 robots and 2 cars.

   Draw all the robots and cars and write the number of batteries in each.

   How many batteries are needed?____

2. Each round cage has 4 parrots and each square cage has 10 canaries. There are 3 round cages and 6 square cages.

   Draw all the cages and write the number of birds in each.

   How many birds are there in all the cages?____

3. A club is going on a trip. Big cars each hold 6 people and there are 12 big cars. Sports cars each hold 2 people and there are 7 sports cars.

   How many people can fit in all the big cars?____

   How many people can fit in all the sports cars?____

   How many people can fit in all the cars together?____

4. Small boxes each have 9 video movies and there are 14 small boxes. Large boxes each have 24 movies and there are 6 large boxes.

   How many movies are there in the small boxes?____

   How many movies are there in the large boxes?____

   How many movies are there in all the boxes?____

5. A club is going for a trip on the river. Each small boat holds 5 people and each large boat holds 15 people. There are 20 small boats and 8 large boats.

   How many people will fit in all the boats?____

6. Each shirt costs $12 and each necktie costs $7.

   What is the total price for 6 shirts and 14 neckties?____

Lesson 7

1. Mrs. Ford baked 5 pans of cookies. Each pan had 4 cookies. She ate 3 cookies.

   How many cookies were left?____

2. A plane has 15 rows of seats. There are 4 seats in each row.

   How many people can fit on the plane?____
   The plane was full. Then it landed and 12 people got off.

   How many people were left?____

3. A classroom has 5 rows of seats. There are 8 seats in each row. The classroom was full. Then 11 children left for the library.

   How many children were left in the room?____

4. Seven planes landed at the Miami airport this afternoon. Each one carried 40 passengers.

   How many passengers were on the planes altogether?____
   A total of 60 passengers got off the planes in Miami.

   How many passengers stayed on the planes and went to other airports?____

5. Five buses stopped at the Cleveland bus station on Tuesday. Each bus had 20 passengers. 10 passengers got off one bus, and 5 got off another bus. The rest stayed on for other cities.

   How many passengers stayed on the buses?____

6. Steve had 20 boxes with 9 wolf spiders in each box. All the spiders from 5 boxes jumped out and attacked Steve, but he crushed them.

   How many spiders were left?____

7. Kim's house has 5 beds. One bed has 4 monsters under it. Another bed has 6 monsters. The other 3 beds each have 11 monsters under them.

   How many monsters are under the beds?____

8. Make up a problem like problem 1.

9. Make up a problem like problem 6.

10. Make up a problem like problem 7.

Lesson 8

In these problems the word PER means FOR EACH.

   Example: *$5 PER shirt means $5 FOR EACH shirt*

1. Evvy bought 7 books at $2 per book.

   How much did she pay altogether?____

2. Joan bought 8 oranges at $.30 per orange.

   How much did she pay altogether?____

3. Cynthia bought 6 apples at $1.00 per apple and 4 bananas at $.50 per banana.

   What was her total bill?____

4. Phil bought 4 shirts at $10 per shirt and 2 belts at $7 per belt.

   How much did he pay altogether?____

5. There are 5 doughnuts per small box and 12 doughnuts per large box.

   How many total doughnuts are there in 10 small boxes and 3 large boxes?____

6. Each toy car requires 3 batteries which cost $2.00 per battery.

   How many batteries are required for 8 cars?____

   How much would it cost to buy batteries for 8 cars?____

7. Twenty children went to a snack shop which charges $3 per sandwich. Eight children each want 2 sandwiches. The other children each want 1 sandwich.

   How much is the total bill?____

Lesson 9

1. Bobby wants to buy a shirt that costs $11. He only has $7.

   How much more money does he need?____

2. Carlos wants to buy a toy car that costs $3 and a boat that costs $4.

   What is the total price for the two toys?____
   Carlos only has $5.

   How much more money does he need to buy the two toys?____

3. Maria wants to buy a skirt for $15 and a blouse for $8. She only has $12.

   How much more money does she need?____

4. Janet wants to buy a glass of milk and 4 cookies. Each cookie costs 50¢ and a glass of milk is 70¢.

   What is the total price for what Janet wants?____
   Janet has $2.

   How much more money does she need to get everything she wants?____

5. Leigh wants 3 pencils and a notebook. The notebook costs $1.20 and each pencil costs 11¢. Ruth only has $1.50.

   How much more money does she need?____

48 MULTIPLICATION

6. Judy wants to buy 2 sodas and 3 doughnuts. A soda costs 40¢ and a doughnut costs 30¢.

   What is the total price for everything Judy wants?____
   Judy has $1.20.

   How much more money does she need?____

7. Sue wants 4 pens that cost 60¢ each and 2 packages of paper that cost $1.10 a package. Sue has $2.50.

   How much more money does she need to buy the pens and pencils?____

8. Jack has $1.05. He wants to buy 2 candy bars and 3 packages of gum. Each candy bar costs 50¢ and each package of gum costs 40¢.

   How much more money does Jack need?____

9. Charles has 3 toy cars that use 5 batteries each. Batteries cost $1 each. Charles has $8.

   How much more money does he need to buy enough batteries for all 3 cars?____

10. When his family eats chicken, Bill will trade 1 leg for 3 wings. Sophie has 4 wings.

    How many more wings does she need to get 2 legs from Bill?____

Lesson 10

If you have trouble answering any of these questions, make a picture to help you see how to get the answer.

1. Paul had 2 jars with 10 mice in each jar. He put 7 mice in John's shirt.

   How many mice did he have left?____

2. Marie had 5 tanks with 8 frogs in each tank. She put 3 frogs in her friend's lunchbox.

   How many frogs did she have left?____

3. Antonio had 7 jars with 20 ants in each jar. He put 11 ants in his parents' bed and 30 ants in his brother's bed.

   How many did he have left?____

4. Andrew baked 5 pans of muffins. Each pan had 6 muffins. While he was watching TV, his dog Rusty ate 8 muffins and his dog Spot ate 3.

   How many were left?____

5. Mike had 5 large cans and 3 small cans. Each large can contained 10 worms and each small can contained 2 worms. He put 7 worms in Aretha's coat.

   How many did he have left?____

6. Write a problem like problem 2.

7. Write a problem like problem 4.

Lesson 11

1. There are 3 tables in each room. Each table accommodates 10 chairs.

   Draw 3 tables and write 10 in each for the chairs. Here is one table.

   How many chairs are there in 1 room?____

   How many chairs are there in 2 rooms?____

   How many chairs are there in 4 rooms?____

50 MULTIPLICATION

2. There are 5 lamps in each room. Each lamp requires 2 lightbulbs.

   Draw 5 lamps and write 2 in each for the bulbs. Here is one lamp.

   How many lightbulbs are there in 1 room?____

   How many lightbulbs are there in 2 rooms?____

   How many lightbulbs are there in 4 rooms?____

3. Each stamp album has 10 pages and there are 4 stamps on each page. Mrs. Brown's class has 7 stamp albums.

   How many stamps are there in 1 album?____

   How many pages are there in 7 albums?____

   How many stamps are there in 7 albums?____

   Can you explain two ways to find the number of stamps in 7 albums?____

4. Each fourth grader has 3 pencils. There are 15 children in each class and there are 5 fourth grade classes in the school.

   How many children are there in 1 fourth grade class?____

   How many pencils are there in 1 fourth grade class?____

   How many fourth grade children are there in the school?____

   How many pencils do all the fourth graders in the school have?____

   Can you explain two ways to find the number of pencils all the fourth graders have?____

5. Each bottle contains 20 vitamin pills and there are 8 bottles in one box. Mr. White bought 7 boxes for his store.

How many pills are there in 1 bottle?____

How many pills are there in 1 box?____

How many bottles are there in 1 box?____

How many bottles are there in 7 boxes?____

How many pills are there in 7 boxes?____

Can you explain two ways to find the number of pills in 7 boxes?____

6. A restaurant orders 6 bags of potatoes a week. Each bag weighs 30 pounds.

How many bags of potatoes does the restaurant buy in 4 weeks?____

How many pounds of potatoes does the restaurant buy in 1 week?____

How many pounds of potatoes does the restaurant buy in 4 weeks?____

Can you explain two ways to find the last answer?____

7. Each student wrote 4 pages with 100 words per page. There are 5 students.

How many words did each student write?____

How many pages did all the students write altogether?____

How many words did all the students write altogether?____

Can you explain two ways to find how many words all the students wrote?____

Lesson 12

1. Each child was given 2 lollipops, and there are 30 children in each class.

How many lollipops were needed for 4 classes of children?____

52 MULTIPLICATION

2. There are 4 cups of soda in a quart of soda. Each box has 2 quart bottles.

   How many cups of soda are there in 5 boxes?____

3. Sally bought 5 bags of flour. Each bag holds 9 pounds and costs $1.10.

   How many pounds of flour did Sally buy?____

   How much did Sally pay for all 5 bags?____

4. Each box of candy costs $4 and contains 20 candies. Ken bought 3 boxes of candy.

   How much did Ken pay for all 3 boxes?____

   How many candies did Ken buy?____

5. Each child sells 2 tickets to a play. There are 21 children in each class and there are 3 classes.

   How many tickets are sold?____

   Explain two ways to get your answer.

6. Bob plays basketball 4 times a month.

   How many times did he play in 3 years?____

   Explain two ways to find your answer.

7. Mr. Bucks owns 5 used car lots and has 30 cars on each lot.

   How many tires must he buy to put new tires on all the cars?____

   Explain two ways to find your answer.

Lesson 13

1. A long distance call costs $2 for the first 3 minutes and 50¢ for each additional minute.

   How much does a 3-minute call cost?_____

   How much does a 4-minute call cost?_____

   How much does a 5-minute call cost?_____

   How much would a 6-minute call cost?_____

   How much would a 7-minute call cost?_____

   How much would a 13-minute call cost?_____

2. A moving company charges $20 for the first 10 pounds and $1 for each additional pound.

   How much does it charge for 11 pounds?_____

   How much does it charge for 15 pounds?_____

   How much does it charge for 22 pounds?_____

3. A postal service charges $10 for the first 4 ounces and $2 for each additional ounce.

   How much would a 7-ounce package cost?_____

4. On the desert, water costs $10 for the first 4 quarts and $5 for each additional quart.

   How much would 8 quarts cost?_____

54 MULTIPLICATION

5. The airport charges $2 an hour for parking up to a maximum of $5 for a day (24 hours).

    How much would it cost to park for 1 hour?____

    How much would it cost to park for 2 hours?____

    How much would it cost to park for 3 hours?____

    How much would it cost to park for 6 hours?____

    How much would it cost to park for 12 hours?____

    How much would it cost to park for 25 hours?____

    How much would it cost to park for 30 hours?____

Lesson 14

1. A small plane has 5 rows of seats. Each row has 2 seats.

    How many seats are there in 1 plane?____

    How many seats are there in 3 planes?____

2. A bus has 20 rows of seats. Each row has 4 seats.

    How many seats are there on 1 bus?____

    How many seats are there on 2 buses?____

3. Mr. Smith owns 6 bike rental stores with 100 bikes in each store. All the bikes need new tires.

    How many tires must he buy?____

4. Ralph's doctor told him to take 3 pills every day for 5 weeks.

    How many pills did he take?____

5. There are 15 pieces of chicken in a box. My family buys 2 boxes a month.

    How many pieces of chicken do we eat in a year?____

MULTIPLICATION 55

6. Bill lives 2 miles from school. He walks to school in the morning and back home in the afternoon.

   How many miles does he walk in 5 school days?____

7. My coach runs 5 miles in 1 hour.

5 miles	5 miles
1 hour	1 hour

   How many miles does she run in 2 hours?____

5 miles	5 miles	5 miles	5 miles
1 hour	1 hour	1 hour	1 hour

   How many miles does she run in 4 hours?____

8. Paula's pony runs 10 miles an hour. This picture shows that in 3 hours the pony runs 30 miles.

10 miles	10 miles	10 miles
1 hour	1 hour	1 hour

   Draw a picture like this on your paper to show how far the pony runs in 4 hours.

   How far does the pony run in 4 hours?____

9. Mrs. Saif drives 50 miles per hour on the highway. This picture shows that in 2 hours she goes 100 miles.

50 miles	50 miles
1 hour	1 hour

   Draw a picture to show how far she goes in 3 hours.

   How far does she go in 3 hours?____

10. A worm crawls 3 feet per hour.

    How far does he crawl in 2 hours?____

    How far does he crawl in 5 hours?____

    How far does he crawl in 50 hours?____

56   MULTIPLICATION

11. A shark swims 3 feet per second.

How far can it swim in 4 seconds?____

How far can it swim in 10 seconds?____

How far can it swim in 1 minute?____

12. A plane travels 300 miles per hour?

How far does it travel in 2 hours?____

How far does it travel in 5 hours?____

Lesson 15

1. There are 4 quarts of milk in 1 gallon. This picture shows there are 8 quarts in 2 gallons.

Draw a picture to show how many quarts there are in 5 gallons.

How many quarts are there in 10 gallons?____

How many quarts are there in 30 gallons?____

2. An airline was warned that the Stowaway Stevens family might sneak on a plane. A flight attendant counted 5 full rows, with 6 people in each row, and an additional 4 people in rows that were not full. The computer said 29 people bought tickets.

How many of the Stowaway Stevens sneaked on the plane?____

3. Coach Davis found 43 baseballs in a locker. Each box he has holds 10 baseballs, and he filled 4 boxes.

How many baseballs were left over?____

4. Lucy baked 3 pans of cupcakes with 10 cupcakes per pan. She bought 2 boxes of cupcake papers which each have 25 papers.

   How many papers were left over?____

5. There are 26 children in Mrs. Braddy's class. Each brought $5 for the party. The cake cost $20, the ice cream cost $30, and the clown act cost $50.

   How much money was left?____

6. Monster had 4 jars with 20 rat tails in each jar. He lost 15 tails and Madman stole 25. But, Weirdo gave Monster 30 tails from his rat ranch.

   Then how many total tails did Monster have?____

7. There are 20 children in the Kiddo Club. Each child had 2 jars, with 5 gumballs in each jar. For Dr. Whimbey's birthday, the club gave him 35 gumballs, which he enjoyed for lunch.

   How many gumballs does the club have left?____

2 Multiplication: Additional Exercises

Lesson 1x

1. A motorcycle has 2 wheels.

   How many wheels do 3 motorcycles have?____

   Do this multiplication on your paper.   2
                                                              ×3

2. A car has 4 wheels.

   How many wheels do 2 cars have?____

   Do this multiplication on your paper.   4
                                                              ×2

3. Calvin has 4 fish tanks. Each tank has 2 fish.

   Draw the tanks with the fish. Here is one tank.

   How many fish does Calvin have altogether?____

   Do this multiplication on your paper.   2
                                                              ×4

60 ADDITIONAL EXERCISES

4. Pete bought 2 apples. Each apple has 3 worms.

   Draw both apples with their worms. Here is one apple.

   How many worms did Pete get altogether?____

   Do this multiplication on your paper.   3
                                                                         ×2

5. Jane has 4 jars. Each jar has 3 eyeballs.

   Draw all the jars with the eyeballs. Here is 1 jar.

   How many eyeballs does Jane have altogether?____

   Do this multiplication on your paper.   3
                                                                         ×4

6. There are 5 Christmas stockings. Each stocking has 3 rings.

   Draw all the stockings with the rings. Here is 1 stocking.

7. A small park has 6 trees. Two squirrels live in each tree.

   How many squirrels are there altogether?____

Lesson 2x

1. A spider has 8 legs.

   How many legs do 3 spiders have?____

   How many legs do 10 spiders have?____

2. A monster has 3 heads. Each head has 7 teeth.

   How many teeth does the monster have altogether?____

3. One cage has 3 dragons. Another cage has 6 dragons.

   How many dragons are there altogether?____

4. Bobby has 3 boxes. Each box has 6 video movies.

   How many movies does he have altogether?____

5. One nickel equals 5 pennies.

   How many pennies equal 2 nickels?____

   How many pennies equal 6 nickels?____

   How many pennies equal 7 nickels?____

6. Les bought a boat for $2 and a toy car for $4.

   How much did he spend?____

7. Each set of doll clothes cost $2. Eve bought 4 sets of doll clothes.

   How much did she spend?____

8. Ian has 6 bags. There are 4 crickets in each bag.

   How many crickets does he have altogether?____

Lesson 3x

1. Wilbert mailed 2 packages on Monday and 5 packages on Tuesday.

   How many packages did he mail?____

2. Helen mailed 2 packages everyday for 5 days.

   How many packages did she mail?____

3. Elsie mailed 5 small packages each costing $2 and 1 large package costing $6.

   How much did it cost to mail all the packages?____

4. You were hungry, so you ordered 20 chocolate ants, 10 fried worms, and a lobster. The ants cost $1 each, the worms cost $3 each, and the lobster cost $7.

   What was the total price of the meal?____

5. A lady walked 4 miles in 1 hour.

   How far did she walk in 2 hours?____

   How far did she walk in 7 hours?____

6. An ant crawled 3 feet an hour.

   How far did he crawl in 2 hours?____

   How far did he crawl in 5 hours?____

7. Mr. Neat gets a haircut every month. He pays $5 for each haircut.

   How much does he spend for haircuts in a year?____

Lesson 4x

1. A pet store has 5 small cages that have 3 birds each. It also has 2 big cages with 10 birds each.

   How many birds are in all the small cages?____

   How many birds are in all the big cages?____

   How many birds does the store have altogether?____

2. A company has 2 trucks. One truck uses 20 gallons of gas 4 times a week. The other uses 30 gallons of gas 2 times a week.

   How much gas do the two trucks use in a week?____

3. The class wants to sell 70 tickets for the school play. Fifteen children each sell 3 tickets.

   How many tickets are left to be sold?____

4. The Rockets basketball team plays 3 games each week in the season. The season is 8 weeks long. Bobby played in all the games except 5 of them.

   In how many games did Bobby play?____

5. Brian bought 5 boxes of apples. Each box weighed 11 pounds. Fifteen pounds of apples were used to make applesauce.

   How many pounds of apples were left?____

6. Mr. and Mrs. Turtle and their twins, Slowpoke and Slowfolk, went on a trip. They packed 10 quarts of mud for each adult and 5 quarts of mud for each child.

   How many quarts of mud did they pack?____

7. A room showing a movie has 12 rows with 6 seats in each row. Andy sold 21 tickets to the movie and Lisa sold 19 tickets.

   How many seats are left?____

Lesson 5x

1. Garcia's class has 5 rows with 8 desks in each row. There are 24 children in class today.

   How many seats are empty?____

2. Mr. Prince bought 4 shirts that cost $12 each and 2 ties costing $5 each. He gave the clerk $100.

   How much change did he get?____

64   ADDITIONAL EXERCISES

3. Yellow boxes hold 6 cupcakes and blue boxes hold 2 cupcakes. Mrs. Clark baked 31 cupcakes. She filled 3 yellow boxes and 4 blue boxes.

   How many cupcakes were left over?____

4. In the fourth grade, 15 children each sold 3 tickets for the school play. In the fifth grade, 12 children each sold 4 tickets.

   How many tickets did the children sell?____

5. A regular jet can carry 210 people. A jumbo jet can carry 515 people.

   How many people can 3 regular jets carry?____

6. Mickey spent 8 dimes and 3 nickels in soda and snack machines.

   How much money did he spend altogether?____

7. Doughnuts cost 30¢ and sandwiches cost $1.

   How much did Jim pay for 2 doughnuts and 4 sandwiches?____

Lesson 6x

1. Herb can mow 2 lawns a day. His big brother Mel can mow 6 lawns a day.

   How many lawns can the boys together mow in a day?____

   How many lawns can the boys together mow in 5 days?____

2. Four boys and 3 girls voted to see a horror movie. But 6 times that many children voted to see a comedy.

   How many children voted to see a comedy?____

3. Coach Davis had 43 baseballs. He filled 4 boxes with baseballs. Each box held 10 baseballs.

   How many were left over?____

4. Cynthia had 30 pounds of peanuts. She filled 2 large cans and 4 small cans with peanuts. Each large can held 6 pounds of nuts and each small can held 3 pounds.

   How many pounds were left over? ____

5. A shelf has 3 pies. Another shelf has 5 pies.

   How many pies are there? ____

6. There are 5 shelves. Each shelf has 3 pies.

   How many pies are there? ____

7. Each large puzzle has 100 pieces and each small puzzle has 50 pieces. Judy has 3 large puzzles and 4 small puzzles. Her baby brother put the pieces from all the puzzles into one big box.

   How many pieces are in the box? ____

8. Jack and Margie took pictures. Jack used 4 rolls of film with 20 pictures on each roll. Margie used 5 rolls of film with 12 pictures on each roll.

   Which child took more pictures? ____

Lesson 7x

1. A bicycle has 2 wheels.

   How many wheels do 10 bicycles have? ____

   How many wheels do 30 bicycles have? ____

2. Mr. Carpenter earns $10 an hour. On Monday he worked 7 hours.

   How much did he make? ____

3. Mr. Plumber earns $10 an hour on weekdays and $15 an hour on Saturdays. He worked 5 hours on Thursday, 7 hours on Friday, and 3 hours on Saturday.

   How much did he make in the 3 days? ____

66 ADDITIONAL EXERCISES

4. What number is 5 more than 19?____

   What number multiplied by 3 is 5 more than 19?____

5. What number is 7 less than 43?____

   What number multiplied by 4 is 7 less than 43?____

6. What number multiplied by 5 is 3 more than 37?____

7. A can of soda in one store cost 50¢. In another store a can of soda is 10¢ cheaper.

   How much do 3 cans of soda cost at the cheaper store?____

Lesson 8x

1. A camp has 5 boats that can each hold 6 children and 10 small boats that can each hold 3 children.

   How many children can the boats hold?____

2. Joe runs 6 miles a week. Ken runs 3 times as many miles a week as Joe does.

   How many miles does Ken run in 4 weeks?____

3. Each player on a baseball team owns 3 uniforms. There are 9 players on each team.

   How many uniforms do 4 teams own?____

4. Eric drew 5 funny faces on his wall every day (including weekends) for 2 weeks.

   How many total faces did he draw?____

5. Goats got into a store and ate 5 shirts and 3 coats. Each shirt cost $7 and each coat cost $15.

   What was the total cost of the clothes eaten by the goats?____

6. Larry and Eddie are saving their money for a video game. Larry has 3 dollars, 2 quarters, and 8 nickels. Eddie has 5 quarters and 4 dimes.

   How much money do they have altogether?____

7. Jack earned $6 a day for 5 days.

   If a radio he wants costs $42, how much more money does he need?____

Lesson 9x

1. Each ice cream cone has 2 scoops of ice cream.

   How many scoops do 12 cones have?____

2. A taxi company owns 9 taxis and wants to replace the tires on all of them.

   How many tires should it buy?____

3. A box of cookies weighs 2 pounds and costs $4.

   How much would 3 boxes of cookies weigh?____

   How much would 30 boxes of cookies weigh?____

   How much would 3 boxes of cookies cost?____

   How much would 30 boxes of cookies cost?____

4. Alice pushes shopping carts for old people and earns 5¢ for each block she pushes a cart.

   How much does she earn for pushing a shopping cart 2 blocks?____

   How much for 3 blocks?____

   How much for 4 blocks?____

   How much for 6 blocks?____

   How much for 8 blocks?____

   How much for 10 blocks?____

   How much for 11 blocks?____

5. One dime equals 10 pennies.

   How many pennies equal 2 dimes?____

   How many pennies equal 3 dimes?____

   How many pennies equal 6 dimes?____

6. One foot equals 12 inches.

12 inches	12 inches	12 inches
1 foot	1 foot	1 foot

   How many inches equal 2 feet?____

   How many inches equal 3 feet?____

   How many inches equal 4 feet?____

   How many inches equal 8 feet?____

Lesson 10x

1. Paul gets 20¢ from the tooth fairy for each tooth. He used his Halloween teeth to fake out the tooth fairy. He put 3 teeth from his mouth and 5 Halloween teeth under his pillow.

   How much money did the tooth fairy bring him?____

2. Bill's car can go 20 miles on 1 gallon of gas.

   How far can it go on 2 gallons of gas?____

   How far on 3 gallons?____

   How far on 4 gallons?____

   How far on 8 gallons?____

3. Jack lost his bus money. He has to walk home 11 miles. He can walk 4 miles an hour. After 2 hours, he sits down to rest.

   How far must he still walk?____

4. Mr. Potbelly bought 5 sandwiches at $1.10 each and 3 sodas at 60¢ each. He gave the clerk $10.

   How much change did he receive?____

5. A concert ticket cost $5.

   How much does it cost for Evvy and her 5 friends to go to the concert?____

6. Each child in the fourth grade needed an injection to prevent polio. There are 5 classes and 24 children in each class. After the doctor gave 83 injections, he ran out of medicine.

   How many children still needed injections?____

7. Each girl in class received 2 teddy bears and each boy received 3 rubber snakes. There are 11 girls and 12 boys in class.

   How many toy animals did the children receive altogether?____

8. Judy can make 10 rings a day. She earns $4 a ring and works every day except Sunday.

   How much does she earn in a month (4 weeks)?____

3 Division

Lesson 1

1. Sue Strangebody baked 6 spiders.

   Draw the 6 spiders on your paper.

   Here is one. 🕷

   Sue put 2 spiders in each bowl.

   Draw a bowl around each group of 2 spiders on your paper. The first bowl is drawn here.

   How many bowls did Sue use?____

   Do this division on your paper. 2)6

2. Mrs. Hawk caught 12 fish.

   Draw 12 fish in groups of 3 like this.

   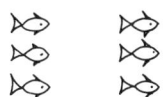

   Mrs. Hawk fed 3 fish to each baby hawk.

   Draw a baby around each group of 3 fish like this.

   How many babies did Mrs. Hawk feed?____

   Do this division on your paper. 3)12

72 DIVISION

3. The pet store received 20 bunnies.

   Draw 20 bunnies in groups of 4 like this.

   The store put 4 bunnies in each cage.

   Draw a cage around each group of 4 bunnies.

   How many groups of 4 bunnies are there in 20 bunnies?____

   Do this division on your paper. 4)‾20

4. Mr. Badtree had 10 worms. He put 2 worms in each apple.

   How many apples did he put worms into?____

   Do the division which gives this answer on your paper.

5. Mr. Sea has 15 shells. He put 3 shells in each box.

   How many boxes did he put shells into?____
   Do the division which gives this answer on your paper.

6. Lisa had 24 tootsie pops. She put 4 pops in each Christmas stocking.

   How many stockings did she put pops into?____
   Do the division on your paper.

7. Miss Chif had 18 fleas. She put 3 on each cat.

   How many cats each got 3 fleas?____
   Do the division on your paper.

8. The Chewmore company puts 5 sticks of gum in each package.

   How many packages can they make from 30 sticks of gum?____

9. Do not divide to conquer.

   Referring to the pictures in problem 1, how many legs do 10 spiders have altogether?____

10. Write a problem like problem 7 or 8.

Lesson 2

1. Billy found 12 dragon eggs. He put 3 eggs in each diamond box.

   How many boxes did he use?____

   Draw all the boxes and show the 3 eggs in each. Here is one.

2. Fifteen children are going on a trip. Each car can take 3 children.

   Draw the 15 children with 3 in each car like this one.

   How many cars are needed?____

   Do this division on your paper. 3)15

3. 63 children are going on a trip. Each car can take 3 children.

   How many cars are needed?____

   Check your answer by multiplying it by 3.

74 DIVISION

4. While Bobby was out walking, 28 dimes suddenly fell from the sky. He gave 4 dimes to each friend.

   How many friends got dimes?____

   Draw the 28 dimes in piles of 4 dimes, one pile for each friend like this.

5. Jean caught 84 baby tigers in India. She gave 4 tigers to each friend.

   How many friends got tigers?____

   Check your answer by multiplying it by 4.

6. Marbles are 5¢ each. Jack has 10¢.

   How many marbles can he buy?____
   Sue has 20¢.

   How many marbles can she buy?____
   Betsy has 80¢.

   How many marbles can she buy?____

7. Betty's favorite record takes 8 minutes to play. She has 24 minutes before leaving for school.

   How many times can she play her record?____

Lesson 3

1. Judy bought 10 angel fish. She put an equal number in each of 2 tanks.

   Draw 2 tanks and divide the fish so an equal number are drawn in each tank.

   Do this division on your paper. 2)10

DIVISION 75

2. Carl caught 40 frogs. He divided them into 2 equal groups and put them into 2 tanks.

   Draw the 2 tanks and write the number of frogs in each tank.

   DO NOT draw all 40 frogs.

   Just write the number of frogs in each tank.

   Do this division on your paper. 2)̄40

3. Ellen had 15 dolls. She divided them into 3 equal groups and put them into 3 boxes.

   Draw 3 boxes and write the number of dolls in each box.

   DO NOT draw all 15 dolls.

   Do this division on your paper. 3)̄15

4. Cynthia collected 45 old soda cans. She divided them into 3 equal piles and put them into 3 bags.

   Draw 3 bags and write the number of cans in each bag.

   Did you do this division on your paper? 3)̄45

5. Sixty-three children visited Disney World. They flew in 3 planes, with an equal number of children in each plane.

   Draw 3 planes and write the number of children in each plane.

   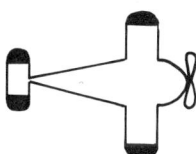

   Did you do this division on your paper? 3)̄63

76 DIVISION

6. Twenty girls went to camp. They slept in 4 tents with an equal number of girls in each tent.

   Draw 4 tents and write the number of girls in each.

   Do this division on your paper. 4)20

7. Sheila has 84 raisins. She wants to divide them equally into 4 groups to bake into 4 giant cookies.

   Draw 4 cookies and write the number of raisins in each.

   Did you do this division on your paper? 4)84

8. Peter Nutter loves peanut butter. He bought 95 jars to hide in 5 cars.

   Draw 5 cars and write the number of jars in each.

   Write the division used to find the number of jars in each car.

Lesson 4

1. Twenty children are going to each lunch. They will sit at 4 round tables.

   Draw 4 round tables and write the number of children at each one.

   Do this division. 4)20

2. Tommy caught 42 lightning bugs. He will put them in 6 jars to light his bedroom.

   Draw 6 jars and write the number of bugs in each.

   Did you do this division? 6)42

3. Jackie had 48 rubber roaches. She put an equal number into each of the 4 beds in her house.

   Draw the 4 beds and write the number of rubber roaches in each.

   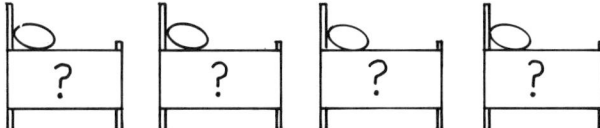

   Did you do this division? 4)̄48

4. Judy found 39 baseball cards. She divided them into 3 equal groups and gave them to her 3 brothers.

   Draw the 3 brothers and write how many cards each received.

   Did you do this division? 3)̄39

5. Bernie got $88 from a witch. He was afraid to keep it. He divided it into 4 equal amounts and gave the money to 4 poor people.

   Draw the 4 people and write the amount of money each received.

   Did you do this division? 4)̄88

6. Fifty-two children were at the aquarium. They saw 4 sharks.

   Draw 4 sharks and write the number of children in each.

   Did you do this division? 4)̄52

78 DIVISION

7. Five children found a pirate's chest with 60 gold coins that they divided equally.

   Draw 5 children and write the number of coins each received.

   Did you do this division? 5)60

8. Eighty-four diamonds were hidden by feeding them to 7 chickens.

   Draw 7 chickens and write the number of diamonds in each.

   Did you do this division? 7)84

Lesson 5

1. The pet store received 18 parrots. It divided them into 3 equal groups and put them in 3 cages.

   How many parrots are in each cage?____

   Multiply your answer by 3 to check it on your paper.

2. Twenty-four children are going on a roller coaster. Four children ride in each coaster car.

   How many cars do they use.

   Multiply your answer by 4 to check it on your paper.

3. Five men bought a total of 30 doughnuts which they divided equally and ate.

   How many doughnuts did each one eat?____
   Multiply your answer by 5 to check it on your paper.

4. Pete caught 48 turtles in a river. He gave 6 turtles to each friend.

   How many friends got turtles?____
   Multiply your answer by 6 to check it on your paper.

5. Sarah picked 42 strawberries. She ate 7 every day.

   For how many days did she eat them?____
   Multiply your answer by 7 to check it on your paper.

6. Fifty-four dogs go to a dog show on an island. They are divided into 6 groups and ride in 6 boats.

   How many dogs go in each boat?____
   Multiply 6 times your answer to check it on your paper.

7. Sue feels 84 tiny legs crawling around in her hair. Each ant has 6 legs.

   How many ants are in Sue's hair?____
   Multiply your answer by 6 to check your answer on your paper.

Lesson 6

1. Each sandwich requires 2 slices of bread. The picture shows 2 slices of bread for one sandwich.

   How many sandwiches can Paul make from 10 slices of bread?____

   How many sandwiches can the chef make from 60 slices of bread?____
   Multiply your answers by 2 to check them on your paper.

80 DIVISION

2. Each club sandwich requires 3 slices of bread. The picture shows 3 slices of bread for one club sandwich.

How many club sandwiches can you make from 60 slices of bread?____

Multiply your answer by 3 to check it on your paper.

3. Polly bought 18 inches of ribbon.

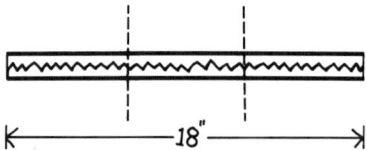

She divided it into 3 equal pieces.

How long is each piece?____

Multiply your answer by 3 to check it on your paper.

4. Donald Doodle made an 84-inch-long noodle. He saw it was wrong and cut it into pieces 6 inches long.

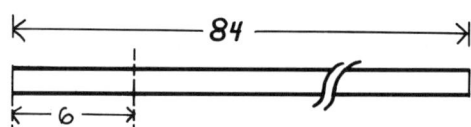

How many 6-inch noodles did he have?____

Multiply your answer by 6 to check it on your paper.

5. Ninety enemy soldiers attacked Super Hero. Super divided them into 6 groups and whipped each group into a pancake.

How many soldiers are in each pancake?____

Multiply your answer by 6 to check it on your paper.

DIVISION 81

6. Monster drank 4 swallows from the soda pond at the soda factory and 60 gallons were gone.

   How many gallons did he drink in each swallow?____

   Multiply your answer by 4 to check it on your paper.

Lesson 7

1. There are 5 cookies in a package. Bobby wants to buy 60 cookies. The picture shows a package with 5 cookies.

   Draw enough packages with 5 in each so the total is 60 cookies.

   Do this division on your paper. 5)60

2. There are 5 cookies in each package.

   How many packages should you buy to get 10 cookies?____

   How many cookies do you get when you buy 3 packages?____

3. There are 10 ice cream pops in each package.

   How many pops do you get when you buy 4 packages?____

   How many packages should you buy to get 60 pops?____

4. Each toy car needs 3 batteries.

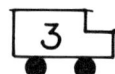 

George bought 12 cars.

How many batteries does he need?____

Tony has 18 batteries.

How many cars can he fill with batteries?____

5. In a restaurant 6 people can sit at each table.

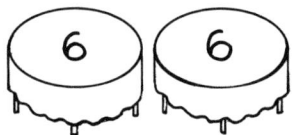

On Monday there were 11 full tables.

How many people were there?____
On Tuesday 54 friends came to the restaurant.

How many tables did they require?____
On Wednesday 132 teachers had a meeting and lunch.

How many tables did they require?____

6. A store offers 1 candy bar free with every 3 you pay for.

Pay 3 3 3 3 3 3
Free 1 1 1 1 1 1

Rosa paid for 12 candy bars.

How many did she get free?____
Joe paid for 60 candy bars.

How many did he get free?____

7. A restaurant lets 1 child eat free with 2 adults who pay for meals.

   A club has 30 adults.

   How many children can eat free with them?____

8. A circus admits 1 adult free with every 4 children that pay.

   Fifty-two children from one school pay to get in.

   How many adults can go with them free?____

9. In 2 cages there are 8 tigers.

   How many tigers are in each cage?____

10. In 3 jars there are 15 snakes.

    How many snakes are in each jar?____

11. The pet store sold 5 kittens for $40. Then it sold 6 puppies for $42.

    Which cost more, 1 kitten or 1 puppy?____
    Bobby bought 2 kittens and 1 puppy.

    How much did he pay?____

84 DIVISION

Lesson 8

1. A store offers 1 shirt free with every 5 paid for. The baseball team paid for 35 shirts.

   How many did they get free?____

2. Jack had 3 frogs and Ann had 15 frogs. They combined their frogs and put an equal number into each of 2 tanks.

   How many frogs are in each tank?____

   Multiply 2 times your answer on your paper to check your division.

3. Twenty girls and 8 boys are going to a picnic. They are divided into 4 equal groups and ride in 4 vans.

   How many children go in each van?____

   Multiply 4 times your answer on your paper to check your division.

4. Mrs. Parker baked 34 cookies. She ate 7 cookies. She divided the rest into 3 equal groups and put them into 3 boxes.

   How many cookies were in each box?____

5. There are 20 slices of bread in each loaf. Sam bought 6 loaves. He made sandwiches using 2 slices in each.

   How many sandwiches did he make?____

6. Eighteen men and 19 women planned to fly to the moon. 2 people got scared and did not go. The rest formed 7 equal groups and went in 7 spaceships.

   How many people flew in each spaceship?____

7. Forty Martians visited our city. Each had 2 heads. They were divided into 5 equal groups and slept in 5 tents.

   How many Martians slept in each tent?____

   How many pillows were needed for each tent?____

8. At the zoo the 24 seals are divided equally into 4 groups and live in 4 pools. Each seal eats 3 fish for lunch.

   How many seals are in each pool?____

   How many fish are put in each pool at lunchtime?____

Lesson 9

1. One can of pigs feet is 40¢.

   How much are 2 cans?____

2. Two cans of monkey tails are 60¢.

   How much is 1 can?____

   Multiply your answer by 2 to check it on your paper.

3. One bottle of bug juice costs 25¢.

   How much are 3 bottles?____

4. Three bottles of banana soda cost 60¢.

   How much is 1 bottle?____

   Multiply your answer by 3 to check it on your paper.

5. Eight bags of ants cost 80¢.

   How much is 1 bag of ants?____

   How much are 4 bags of ants?____

6. Three cans of popcorn cost $3.60.

   How much is 1 can?____

   How much are 2 cans?____

7. Seven bags of bugs cost 70¢.

   How much do 4 bags cost?____

86 DIVISION

8. Eight baby tiger sharks cost $16.

   How much do 3 cost?_____

9. What number divided by 2 equals 3?_____

   What number divided by 2 equals 5?_____

   What number divided by 2 equals 10?_____

10. What number divided by 4 equals 3?_____

    What number divided by 4 equals 5?_____

    What number divided by 4 equals 8?_____

Lesson 10

1. Two perfumed erasers cost 60¢.

   How much does 1 cost?_____

2. Paul paid $5 for a pen and a notebook. The pen cost $2.

   How much did the notebook cost?_____

3. Robert paid 90¢ for a glass of milk and 2 cookies. The milk cost 50¢.

   How much did the 2 cookies together cost?_____

   How much did 1 cookie cost?_____

4. Jim paid $16 for a glove and 2 baseballs. The glove costs $10.

   How much does 1 baseball cost?_____

   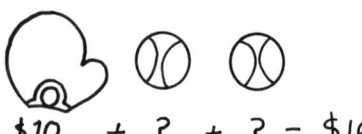

   $10 + ? + ? = $16

DIVISION 87

5. Al paid $32 for a toy boat and 4 toy soldiers. The boat costs $8.

   How much does 1 soldier cost?____

   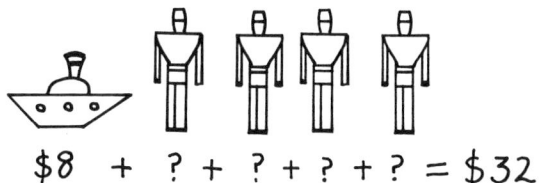

   $8 + ? + ? + ? + ? = $32

6. Madman paid $54 for 5 bottles of blood and a lizard. The lizard costs $4.

   How much is 1 bottle of blood?

   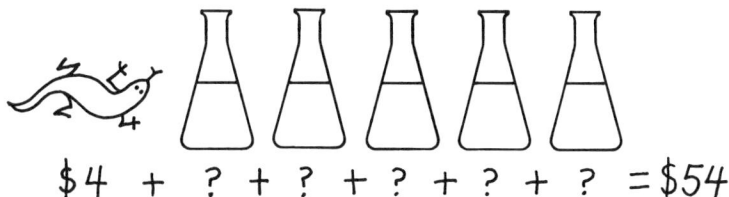

   $4 + ? + ? + ? + ? + ? = $54

7. What number divided by 3 equals 5?____

8. What number is 2 more than 3?____

   What number divided by 4 is 2 more than 3?____

9. What number is 5 less than 11?____

   What number divided by 3 is 5 less than 11?____

10. What number divided by 2 is 5 more than 9?____

Lesson 11

1. Bob bought 3 stamps for 39¢.

   How much was 1 stamp?____

   [?] + [?] + [?] = 39¢

88 DIVISION

2. Kim paid $23 for a doll and 3 doll dresses. The doll costs $11.

   How much was each dress?____

   $11 + ? + ? + ? = $23

3. Kathy paid $17 for 2 fish and a dog. Each fish costs $5.

   How much did the 2 fish together cost?____

   How much did the dog cost?____

4. Monster paid $75 for 2 spiders and a snake. Each spider cost $30.

   How much did the 2 spiders together cost?____

   How much did the snake cost?____

5. Bobby paid $22 for 2 trucks and 3 toy planes. Each truck cost $5.

   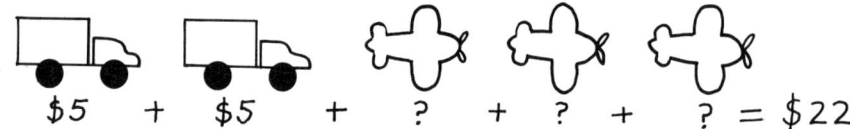
   $5 + $5 + ? + ? + ? = $22

   How much did the 2 trucks together cost?____

   How much did the 3 planes together cost?____

   How much did 1 plane cost?____

6. Easter Bunny paid $38 for 3 chocolate eggs and 4 bags of jelly beans. Each chocolate egg cost $2.

   How much did all the eggs together cost?____

   How much did 1 bag of jelly beans cost?____

DIVISION 89

7. Three sandwiches and 6 orange drinks cost $6.30. Each sandwich cost $1.10.

   How much was each orange drink?____

Lesson 12

1. Shana paid $5.30 for a lemon pie and 3 plastic worms to hide in it. The pie cost $2.

   How much did each worm cost?____

   $2 + ? + ? + ? = $5.30

2. Dragonman paid $55 for 2 pigs and 1 elephant. Each pig cost $15.

   How much did the elephant cost?____

   $15 + $15 + ? = $55

3. Mary paid $3.60 for 3 doughnuts and 2 cups of coffee. Each doughnut costs $1.

   How much does 1 cup of coffee cost?____

   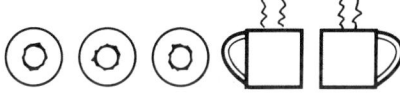

4. Daniel paid $11.60 for 4 water pistols and 3 toy police badges. Each pistol cost $2.

   How much did each badge cost?____

5. Joker Jake paid $10.90 for 3 cans of fake fried eggs, 4 rubber snakes, and 2 rubber spiders. Each rubber snake cost $1 and each rubber spider cost $1.50.

   How much is 1 can of fake eggs?____

6. Eight men, 7 women, and 24 children planned to go on a picnic. But 6 children and 3 men got sick, so they could not go. The other people drove to the picnic with 5 people in each car.

   How many cars did they use?____

   Multiply your answer by 5 on your paper to check your division.

DIVISION

7. A Martian has 5 heads and many arms. In a room with several Martians, there are 20 heads and 36 arms.

   How many arms does 1 Martian have?____

8. Make up a problem like problem 1 and let another student solve it.

Lesson 13

1. A store sold 30 cans of soda. It buys the soda in boxes, with 6 cans in each box.

   How many boxes of soda were used?____

   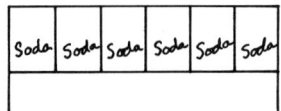

   Multiply your answer by 6 to check it on your paper.

2. The cafeteria sold 15 hotdogs on Monday and 17 hotdogs on Tuesday. It buys the hotdogs in boxes, with 8 hotdogs per box.

   How many boxes of hotdogs were used in the 2 days?____

   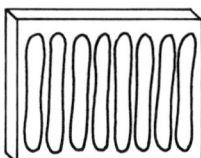

3. A restaurant sold 32 doughnuts on Monday, 30 doughnuts on Tuesday, and 28 doughnuts on Wednesday. They buy the doughnuts in boxes, with 10 doughnuts per box.

   How many boxes of doughnuts did they use in the three days?____

4. A store sold 15 shirts one day. It receives the shirts in boxes with 5 shirts per box. Each box costs $13.

   How much did it pay for the shirts it sold?____

DIVISION 91

5. Carver Elementary School used 50 sheets of paper on Monday, 150 sheets on Tuesday, 200 sheets on Wednesday, and 100 sheets on Thursday. Paper costs $1 for 100 sheets.

   What was the total cost of the paper for the four days?____

6. A store sold 6 turtles. It makes $5 for every 2 turtles it sells.

   How much money did it make?____

   How much did the store make on a day it sold 22 turtles?____

7. A store sold 48 puppies. It makes $7 for every 3 puppies it sells.

   How much did it make?____

8. A flower seller on the street sold 15 pink roses, 20 red roses, and 7 white roses. She makes $3 for every 2 roses sold.

   How much did she make?____

9. A restaurant sold 8 pies on Monday, 3 pies on Tuesday, 9 pies on Wednesday, and 10 pies on Thursday. It makes $21 for every 5 pies sold.

   How much did it make in the 4 days?____

10. A store charges $1 per cookie but gives 1 free cookie with every 5 paid for.

    How many cookies could you get for $20?____

Lesson 14

1. In June Mr. Racket bought 5 boxes of tennis balls which contained a total of 60 balls. In July he bought 3 boxes.

   How many tennis balls did he buy in July?____

92 DIVISION

2. Mr. Jake had 72 tickets to sell. He gave 30 to Lucy and the rest to Pee Wee. Pee Wee sold an equal number of tickets to each of 7 friends.

   How many tickets did each friend buy?____

3. Paul and Joan bought 1 pound of meat and made 2 equal size hamburgers.

   How many ounces of meat did each get?____ Remember that 1 pound equals 16 ounces.

4. Four girls bought a 2-foot long piece of ribbon and cut it into 4 equal pieces.

   How many inches long is each piece?____ Remember that 1 foot equal 12 inches.

5. Larry needs 5 lemons and 5 limes to make 1 quart of punch.

   How many quarts can he make with 10 lemons and 10 limes?____

   How many quarts can he make with 20 lemons and 20 limes?____
   How many quarts can he make with 5 lemons and 6 limes? answer: 1

   How many quarts can he make with 5 lemons and 8 limes?____
   How many quarts can he make with 12 lemons and 10 limes? answer: 2

   How many quarts can he make with 10 lemons and 13 limes?____

   How many quarts can he make with 15 lemons and 17 limes?____

   How many quarts can he make with 15 lemons and 83 limes?____

   How many quarts can he make with 90 lemons and 20 limes?____

6. Fifty-four people signed up to go on a tour of a park. One guide can take a group of 6 people.

   How many guides are needed?____

7. Eight cars arrived at the zoo. Each car had 5 children. They separated into groups of 10.

   How many groups were there?____

DIVISION 93

8. Jars of paint cost $2 each. Four children each bought 6 jars of paint. They divided the cost equally.

   How much did each child pay?____

9. Three families gave a total of $6 to the Red Cross each month for the whole year.

   How much did each family give in a year?____

10. Hats are packed 5 to a box. The Hat Store paid $30 for 2 boxes of hats.

    How much did each hat cost?____

Lesson 15

1. Each box of balloons costs $3. Julie bought several boxes of balloons. She gave the clerk $20 and got $2 change.

   How many boxes of balloons did she buy?____

2. There are 20 children in each class and there are 6 classes. Two children sit at each desk.

   How many desks are there in the 6 classrooms?____

3. Donna bought 3 boxes of baseball caps for $60. Each box has 10 caps.

   How much does 1 cap cost?____

4. The regular price for a box of 2 dolls is $30. On a special sale the box of 2 dolls costs only $16.

   How much money do you save on each doll?____

5. Thirty-five students rented a bus for $210 to go to a football game in another town. Five students decided not to go.

   How much did each of the other students have to pay for the bus?____

94 DIVISION

6. Bill ate some worms that made him sick for 56 days.

   For how many weeks was Bill sick?____

7. Sheila is paid $9 by the pet store for every 4 lizards she catches. In January she caught 44 lizards.

   How much did she make?____

8. Six raisins and 2 walnuts are used in each cookie. Bob and Wanda have 42 raisins.

   How many walnuts do they need to bake cookies with these raisins?____

9. There are 51 guests at a cowboy motel. Four guests at a time can be photographed on horses. Three guests are afraid of horses, but the rest want to be photographed.

   How many photographs must be taken?____

10. A teacher made a piece of red licorice that was 71 inches long. She cut off 3 pieces that were each 2 inches for herself. She cut the rest into 5-inch pieces for her students.

    How many pieces did she have for students?____

Lesson 16

1. Paula baked 40 cookies. First she filled 3 green boxes which held 10 cookies each.

   How many cookies did she put into green boxes?____

   How many cookies were left over after the green boxes were filled?____
   Paula used the leftover cookies to fill red boxes which only held 5 cookies each.

   How many red boxes did she fill?____

2. Jack had 100 marbles. First he filled 3 black bags which each held 20 marbles.

   How many marbles did he put into black bags?____

   How many marbles were left over after the black bags were filled?____
   Jack used the leftover marbles to fill white bags which only held 5 marbles each.

   How many white bags did he fill?____

3. The pet store received 38 birds. First, 3 yellow cages which held 10 birds each were filled. The rest were put into blue cages which only held 2 birds each.

   How many blue cages were used?____

4. Al made 20 quarts of chocolate milk for a party. First he filled 4 bottles that held 3 quarts each. He used the rest to fill smaller bottles which only held 2 quarts each.

   How many smaller bottles did he fill?____

5. Courtney made 72 quarts of lemonade for the school picnic. First she filled 10 bottles that held 6 quarts each. She used the rest to fill smaller bottles which only held 4 quarts each.

   How many smaller bottles did she fill?____

3 Division: Additional Exercises

Lesson 1x

1. Mrs. Jones baked 12 cookies. She put 4 cookies in each box.

   How many boxes did she fill? ____

   ●● ●● ●●
   ●● ●● ●●

   Do this division: 4)12

2. Tom bought 10 fish. He put 2 fish in each tank.

   Draw enough tanks with 2 fish in each so the total is 10 fish.

   Do this division: 2)10

3. Bobby had 20 marbles. He gave 4 marbles to each of his friends. The marbles he gave to 2 friends are shown.

   Draw all the friends and marbles on your paper.

   Do this division on your paper: 4)20

98 DIVISION: ADDITIONAL EXERCISES

4. Lenny had 15 pennies. He put 3 pennies in each pile.

Draw enough piles of 3 pennies so the total is 15.

Do this division on your paper: 3)15

5. Mrs. Lewis needs to buy 27 oranges. There are 3 oranges in a package.

How many packages does she need?

Do this division on your paper: 3)27

Do this multiplication to check your answer:
$$\begin{array}{r} 9 \\ \times 3 \\ \hline \end{array}$$

6. Mrs. Dodge needs to buy 48 pencils for the children in her class. Each package contains 4 pencils.

How many packages must she buy?____

7. Billy had 18 white mice. He put 6 mice into each cage.

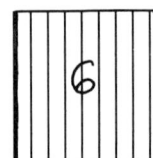

How many cages did he use?____

Do this division on your paper: 6)18

Do this multiplication to check your answer:
$$\begin{array}{r} 6 \\ \times 3 \\ \hline \end{array}$$

… DIVISION: ADDITIONAL EXERCISES 99

Lesson 2x

1. Five frogs weigh 1 pound.

    5 frogs   5 frogs   5 frogs
    1 pound 1 pound 1 pound

    How much do 10 frogs weigh?____

    How much do 15 frogs weigh?____

    How much do 20 frogs weigh?____

    How much do 30 frogs weigh?____

    How much do 40 frogs weigh?____

    How much do 55 frogs weigh?____

    Do this division on your paper 5)40

    Do this multiplication to check your answer.   8
                                                         ×5

    Do this division on your paper: 5)55

2. Thirty children are going to the circus. Six children can go in each van. The picture shows a van with the number 6 to mean 6 children.

    Draw enough vans with 6 in each so the total is 30 children.

    Do this division: 6)30

    Do this multiplication to check your answer.   6
                                                         ×5

3. Twenty-four fish will be fed to sharks. Each shark can eat 4 fish.

    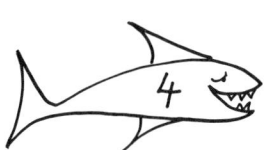

    Draw enough sharks with 4 in each so the total is 24 fish.

    Do this division: 4)24

    Do this multiplication to check your answer.   6
                                                       ×4

100 DIVISION: ADDITIONAL EXERCISES

4. A club sandwich uses 3 slices of bread.

How many club sandwiches can you make with 33 slices of bread?____

Check your answer by multiplying it by 3 on your paper.

5. There are 10 sticks of gum in a package.

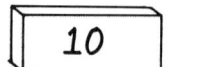 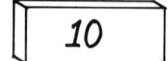

Leon needs 60 sticks of gum for a party.

How many packages of gum should he buy?____

Check your answer by multiplying it by 10 on your paper.

6. The pet store had 25 white rats. Seven rats escaped. The store put the rest into cages, with 3 rats in each cage.

How many cages did they use?____

Check your answer by multiplying it by 3 on your paper.

7. Pauline sold 3 pink roses. 4 red roses, and 1 white rose.

Draw all the roses she sold.

Pauline makes $5 for every 2 roses she sells.

Circle groups of 2 roses on your picture and write $5 like this.

How much did Pauline make altogether?____

DIVISION: ADDITIONAL EXERCISES   101

Lesson 3x

1. Maria has 8 diamond rings. She puts them in 2 boxes. Each box has the same number of rings.

   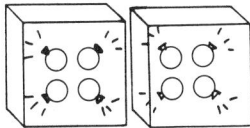

   How many rings are in each box?____

   Do this division on your paper. 2)8̄

2. Jack has 15 baseball cards. He put an equal number into each of 3 boxes.

   How many cards are in each box?____

   Do this division on your paper. 3)15̄

3. Mrs. Smith has a piece of ribbon 8 inches long.

   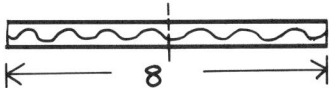

   She wants to cut it into 2 equal pieces.

   How long should each piece be?____

4. Mr. Plumber has a piece of pipe 6 feet long.

   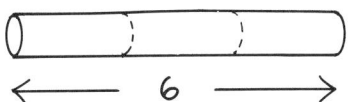

   He wants to cut it into 3 equal pieces.

   How long should each piece be?____

102 DIVISION: ADDITIONAL EXERCISES

5. Mr. Carpenter has a piece of wood 12 inches long.

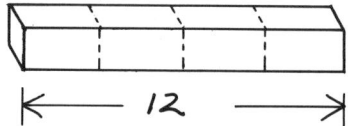

He needs to cut it into 4 equal pieces.

How long should each piece be?____

Do this division on your paper. 4)̄12

6. Thelma had 40 books. She put them on 4 shelves, with an equal number of books on each shelf.

Draw 4 shelves like these on your paper.

Write the number of books Thelma put on each shelf on your paper.

Do this division on your paper. 4)̄40

Lesson 4x

1. Paula has 44 music tapes which she wants to put into 4 gift boxes, with an equal number of tapes in each box.

Draw 4 boxes like these but write the number of tapes in each box instead of a question mark.

?	?
?	?

Do this division on your paper. 4)̄44

Check your answer by multiplying it by 4.

2. The pet store has 21 goldfish. It wants to put them into 3 tanks, with an equal number in each tank.

Draw 3 tanks like these and write the number of goldfish in each tank.

Do this division on your paper. 3)̄21

Check your answer by multiplying it by 3.

3. Phil had 75 worms to go fishing. He kept 35 for himself. He gave the remaining worms to 2 fishing friends.

How many worms did each friend get?____

4. Five children found a pirate's chest with 60 gold coins that they divided equally. Draw 5 children and write the number of coins each received.

Do this division on your paper. 5)60

5. Thirteen girls and 15 boys are going to a picnic. They will go in 4 vans.

Draw 4 vans and write the number of children in each.

6. One pitcher of milk serves 6 people.

    1 pitcher   1 pitcher
    6 people   6 people

How many pitchers are needed for 18 people?____

How many pitchers for 30 people?____

How many pitchers for 66 people?____

7. In 2 boxes there are 8 candy bars.

How many candy bars are in 1 box?____

8. In 5 boxes there are 30 doughnuts.

   How many doughnuts are in 1 box?____

9. In 4 boxes there are 80 balloons.

   How many balloons are in 1 box?____

Lesson 5x

1. Tim has 35 marbles. He put an equal number of marbles into each of 5 bags.

   Draw the 5 bags and write the number of marbles in each.

   Do this division on your paper. 5)35

   Check your answer by multiplying it by 5 on your paper.

2. Thirty-eight people were planning to go on a picnic. Then 8 people decided not to go. The rest went in cars, with 5 people in each car.

   How many cars did they use?____

3. Steve caught 19 grasshoppers. Four got away. With the rest, he put 3 grasshoppers in each bed of several friends.

   How many friends got grasshoppers in their beds?____

4. Write the number to fill each blank.

   4 quarters equal ____ dollar

   8 quarters equal ____ dollars

   12 quarters equal ____ dollars

   24 quarters equal ____ dollars

5. Mr. Plumber had a pipe 45 inches long. He cut off 3 inches. He cut the rest into 6 equal pieces.

   How long was each piece?____

6. Carol uses 4 oranges to make 1 pitcher of orangeade. She has 36 oranges.

   How many pitchers of orangeade can she make?____
   Check your answer by multiplying it by 4 on your paper.

7. There are 24 children in class. They are divided into 8 equal groups to work on math problems.

   How many children are in each group?____
   Check your answer by multiplying it by 8.

Lesson 6x

1. A bike cost $40. Bill paid for it in 4 equal payments.

   How much was each payment?____

2. A doll house cost $110. Lois made a down payment of $30. She paid the rest in 4 equal payments.

   How much was each payment?____

3. Sixty-five people want to go to the moon. A spaceship can only take 5 people each trip.

   How many trips must it make to take all the people?____

4. Larry has been sitting at his desk for 42 minutes. For 10 minutes he read a book on making kites. In the remaining time he made 4 kites.

   How long did it take to make 1 kite?____

5. Paul rode his bike 9 miles in 1 hour.

   How long will it take him to go 27 miles?____

106 DIVISION: ADDITIONAL EXERCISES

6. Five erasers and 1 pencil cost 65¢. The pencil cost 15¢.

   How much did the 5 erasers cost?____

   How much would 1 eraser cost?____

7. What number divided by 4 is 2?____

   Do this division: 4)̄8

8. What number divided by 3 equals 7?____

   Do this division: 3)̄21

9. What number divided by 2 equals 10?____

Lesson 7x

1. Two cans of peas cost $2.

   How much does 1 can cost?____

2. Three cans of tuna cost $3.30.

   How much does 1 can cost?____

3. Four jars of root rot jam cost $4.80.

   How much does 1 jar cost?____

4. Mr. Wrightson paid $14 for a shirt and 2 ties.

   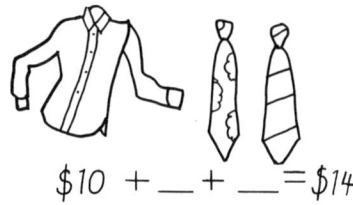

   $10 + ___ + ___ = $14

   How much did 1 tie cost?____

5. Alien paid $19 for a can of toes and 2 jars of eyeballs. The can of toes cost $3.

   How much was each jar of eyeballs?____

6. Katrina paid $17 for a skirt and 2 blouses. The skirt was $7. Each blouse was the same price.

   How much was each blouse?____

7. Bud paid $42 for a baseball glove and 3 bats. The glove cost $12.

   How much was each bat?____

8. A table and 4 chairs cost $110. The table cost $50.

   How much did 1 chair cost?____

9. A witch paid $22 for a broom and 5 spider webs. The broom cost $2.

   How much was 1 spider web?____

4 Fractions

Lesson 1

1. Here are 3 balls. $\frac{1}{3}$ of the balls are black.

   Draw 3 hearts. Make $\frac{1}{3}$ of the hearts black.

2. Draw 4 balls. Make $\frac{1}{4}$ of the balls black.

3. Here are 4 cookies. $\frac{3}{4}$ of the cookies are chocolate.

   What fraction of the cookies is vanilla?____

4. Here are four spiders. One is black and 3 are yellow.

   What fraction of the spiders is black?____

   Did you answer $\frac{1}{4}$ for the last question?

   What fraction of the spiders is yellow?____

   Did you answer $\frac{3}{4}$ for the last question?

110 FRACTIONS

5. Here are 3 marbles. Two are black and 1 is white.

What fraction of the marbles is black?____

Did you answer $\frac{2}{3}$ for the last question?

What fraction of the marbles is white?____

6. Here are 5 candies. Three are chocolate and 2 are butterscotch.

What fraction of the candies is chocolate?____

What fraction of the candies is not chocolate?____

What fraction of the candies is butterscotch?____

7. Here are 6 balloons. Two are blue and the rest are white.

What fraction of the balloons is blue?____ Reduce to lowest terms____

What fraction of the balloons is not blue?____ Lowest terms____

What fraction of the balloons is white?____ Lowest terms____

FRACTIONS 111

8. Here are 9 pens. Some are black and the others are white.

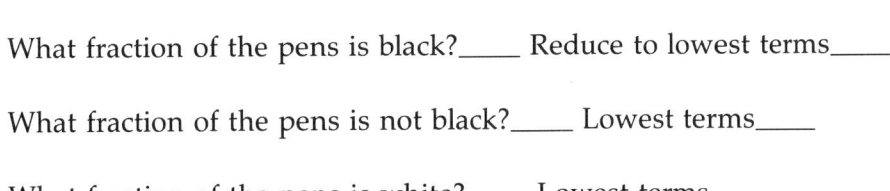

What fraction of the pens is black?____ Reduce to lowest terms____

What fraction of the pens is not black?____ Lowest terms____

What fraction of the pens is white?____ Lowest terms____

9. Here are 8 cupcakes. Some are chocolate and others are vanilla.

What fraction of the cupcakes is chocolate?____

What fraction of the cupcakes is not chocolate?____

10. These spiders were in Barb's bed. The poisonous spiders are marked with a X.

What fraction of the spiders is poisonous?____

What fraction of the spiders is not poisonous?____

11. Wilson caught 3 grasshoppers. He kept 1 and let the others go.

What fraction of the grasshoppers did he keep?____

What fraction of the grasshoppers did he let go?____

12. Darren has 10 cats. He gave 2 of them to his cousin.

What fraction of his cats did Darren give his cousin?____ Lowest terms____

What fraction of the cats did Darren keep?____ Lowest terms____

112 FRACTIONS

13. Shana picked 40 flowers. She sold 10 of them and kept the rest.

What fraction of the flowers did Shana sell?____ Lowest terms____

What fraction of the flowers did Shana keep?____ Lowest terms____

Lesson 2

1. Here are some hearts and diamonds.

What fraction of the figures is hearts?____ Lowest terms____

What fraction of the figures is not hearts?____ Lowest terms____

What fraction of the figures is diamonds?____

2. Here are some cookies shaped like stars, circles, and triangles.

What fraction of the cookies is stars?____ Lowest terms____

What fraction of the cookies is circles?____ Lowest terms____

What fraction of the cookies is triangles?____ Lowest terms____

3. Here are some circles and squares.

What fraction of the figures is circles?____ Lowest terms____

What fraction of the figures is squares?____ Lowest terms____

What fraction of the figures is black?____ Lowest terms____

Did you answer $\frac{1}{3}$ for the last question?____

What fraction of the figures is white?____ Lowest terms____

4. Here are some chocolate and vanilla candies shaped as hearts, stars, and Christmas trees.

What fraction of the candies is chocolate?____

What fraction of the candies is vanilla?____

What fraction of the candies is hearts?____

What fraction of the candies is stars?____

What fraction of the candies is trees?____

5. Weirdo had 6 fish, 4 snakes, and 8 rats in his lunchbox.

What fraction of the items were fish?____ Lowest terms____

What fraction of the items were snake?____ Lowest terms____

What fraction of the items were rat?____ Lowest terms____

6. Chris cooked 13 cows and invited Carrie for dinner. Chris ate 5 cows and Carrie ate the rest.

What fraction of the cows did Chris eat?____

What fraction of the cows did Carrie eat?____

7. Twelve monsters are at the door. Four monsters are green, 6 monsters are purple, and the rest are blue.

How many monsters are blue?____ Lowest terms____

What fraction of the monsters is blue?____ Lowest terms____

What fraction of the monsters is not blue?____ Lowest terms____

What fraction is not green?____ Lowest terms____

114   FRACTIONS

8. Lori had 6 stickers. She gave $\frac{1}{2}$ of them to Bessy.

   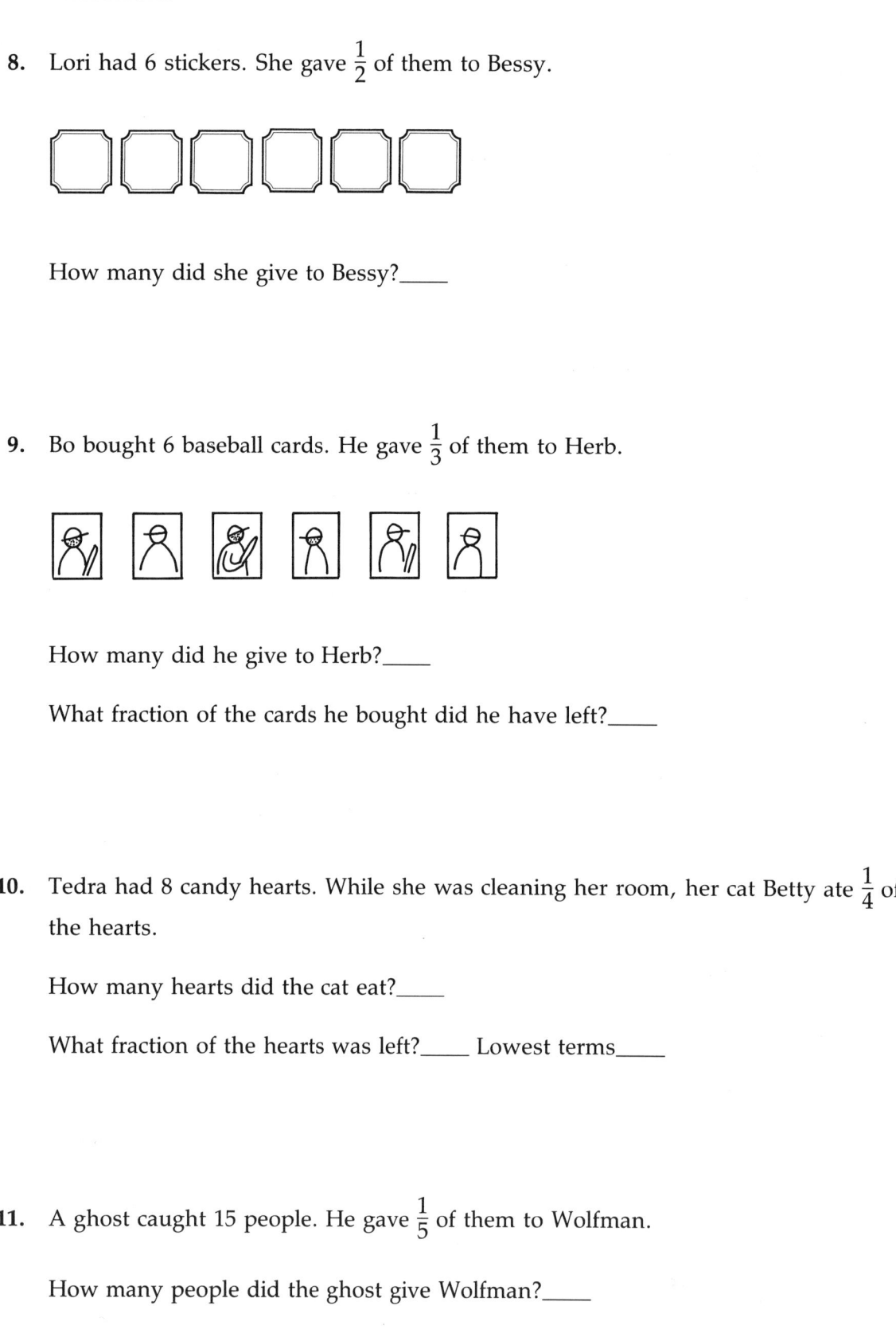

   How many did she give to Bessy?____

9. Bo bought 6 baseball cards. He gave $\frac{1}{3}$ of them to Herb.

   How many did he give to Herb?____

   What fraction of the cards he bought did he have left?____

10. Tedra had 8 candy hearts. While she was cleaning her room, her cat Betty ate $\frac{1}{4}$ of the hearts.

    How many hearts did the cat eat?____

    What fraction of the hearts was left?____ Lowest terms____

11. A ghost caught 15 people. He gave $\frac{1}{5}$ of them to Wolfman.

    How many people did the ghost give Wolfman?____

    What fraction of the people he caught did the ghost have left?____ Lowest terms____

Lesson 3

1. Here are 9 figures.

   What fraction of the figures is circles?____

   What fraction of the figures is squares?____

   What fraction of the figures is black?____

   Do this addition: $\frac{2}{9} + \frac{4}{9} =$

2. Jay-Jay has 2 giant goldfish, 4 angle fish, and 1 dog.

   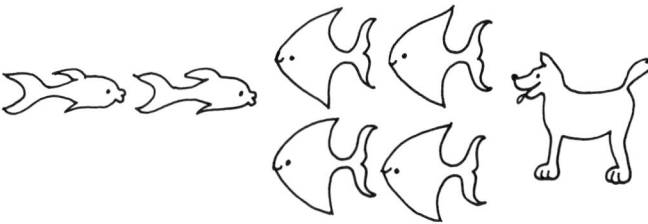

   What fraction of his pets is goldfish?____

   What fraction of his pets is angle fish?____

   What fraction of his pets is fish (not dogs)?____

   Do this addition: $\frac{2}{7} + \frac{4}{7} =$

3. Kathy bought 1 orange drink, 3 grape drinks, and 6 Cokes.

   How many total drinks did she buy?____

   What fraction of the drinks were orange?____

   What fraction of the drinks were grape?____

   What fraction of the drinks were fruit drinks?____

   Do this addition: $\frac{1}{10} + \frac{3}{10} =$

FRACTIONS

4.

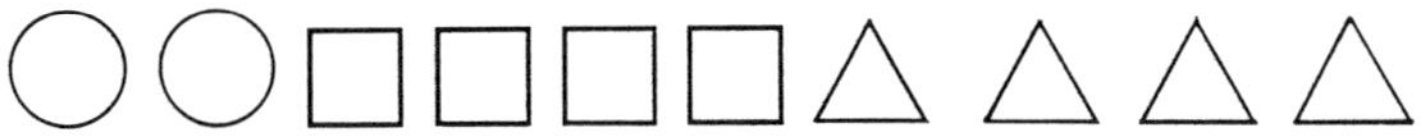

○ ○ □ □ □ △ △ △ △

What fraction of the figures is circles?____

What fraction of the figures is squares?____

What fraction of the figures is *not* triangles?____

Show how to get this answer by adding the first two answers.

5. Trina had 7 Terrible Toad toes in a jar. She took 2 toes on a trip for luck and Tony took 3 toes.

What fraction of all the toes did Trina take?____

What fraction of all the toes did Tony take?____

What fraction of all the toes did Trina & Tony together take?____

Show how to get this answer by adding fractions.

6. Keith had 19 flies. His frog Croker ate 6 flies and his frog Hoppy ate 5 flies.

What fraction of all the flies did Croker eat?____

What fraction of all the flies did Hoppy eat?____

What fraction of all the flies did the 2 frogs together eat?____

Show how to get this answer by adding fractions.

7.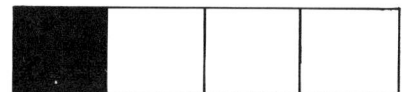

What fraction of the box is black?____

What fraction of the box is white?____

What is the sum of your two answers?____

Do these additions and subtractions: $\frac{1}{4} + \frac{3}{4} =$ ____

$\frac{4}{4} - \frac{1}{4} =$ ____

$\frac{4}{4} - \frac{3}{4} =$ ____

8.

What fraction of the marbles is black?____

What fraction of the marbles is white?____

What is the sum of your two answers?____

Do these subtractions: $\frac{5}{5} - \frac{1}{5} =$

$\frac{5}{5} - \frac{4}{5} =$

9.

What fraction of the marbles is black?____

What fraction of the marbles is white?____

What is the sum of your two answers?____

Do these subtractions: $\frac{5}{5} - \frac{3}{5} =$

$\frac{5}{5} - \frac{2}{5} =$

118 FRACTIONS

10. Seven cookies are in a box. Two cookies are vanilla and the rest are chocolate.

What fraction of the cookies is vanilla?____

What fraction of the cookies is chocolate?____

What is the sum of your two answers?____

Do these subtractions: $\frac{7}{7} - \frac{2}{7} =$

$\frac{7}{7} - \frac{5}{7} =$

Lesson 4

1. There are 5 bugs in a bottle. Three bugs are brown and the rest are red.

What fraction of the bugs is brown?____

What fraction of the bugs is red?____

Add your two answers on your paper.

2. ○○△▢▢▢

What fraction of the figures is circles?____

What fraction of the figures is triangles?____

What fraction of the figures is squares?____

Add your three answers on your paper.

3. There are 4 cookies in a box. One is vanilla and the rest are chocolate.

   What fraction of the cookies is vanilla?____

   What fraction of the cookies is chocolate?____

   Show how to get the second answer by subtracting from $\frac{4}{4}$ the first answer.

   $\frac{4}{4} -$ ____ $=$ ____

4. There are 9 eyeballs in a jar. Two eyeballs are blue and the rest are brown.

   What fraction of the eyeballs is blue?____

   What fraction of the eyeballs is brown?____

   Show how to get the second answer by subtracting the first answer from $\frac{9}{9}$.

   $\frac{9}{9} -$ ____ $=$ ____

5. There are 27 worms in a jar. 11 are red and the rest are gray.

   What fraction of the worms is red?____

   What fraction of the worms is grey?____

   Show how to get the second answer by subtracting the first answer from $\frac{27}{27}$.

   $\frac{27}{27} -$ ____ $=$ ____

6. Alex has 8 cobra snakes, 3 rattle snakes, and 2 poison sea snakes.

   What fraction of the snakes is cobras?____

   What fraction of the snakes is rattle snakes?____

   What fraction of the snakes is the cobras and rattle snakes together?____

   Subtract your last answer from $\frac{13}{13}$. $\frac{13}{13} -$ ____ $=$ ____

   What fraction of the snakes is poison sea snakes?____

120   FRACTIONS

7. Sally ate $\frac{1}{3}$ of her sandwich.

   How much is left?____

8. Paul ate $\frac{1}{4}$ of her candy bar.

   How much is left?____

9. Courtney ate $\frac{1}{5}$ of a pie, and Tom ate $\frac{2}{5}$ of it.

   What fraction of the pie is left?____

   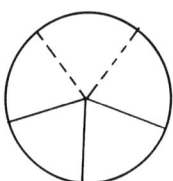

10. Bill ate $\frac{1}{7}$ of a pie and Jack ate $\frac{2}{7}$ of it. What fraction of the pie is left?____

11. Ann mowed $\frac{1}{5}$ of her lawn before lunch and $\frac{1}{5}$ of her lawn after lunch.

    How much of the lawn did she mow?____

    How much of the lawn is not mowed?____

12. Dracula had a full barrel of blood. On Monday night he drank $\frac{1}{11}$ barrel of blood.

    On Tuesday night he had a vampire party and used $\frac{3}{11}$ barrel of blood.

    How much blood is left?____

Lesson 5

1. Here are 4 cookies. $\frac{1}{2}$ of the cookies are chocolate.

   How many cookies are chocolate?____

   Fill the blank: $\frac{1}{2}$ of 4 = ____

   Do this division on your paper: $2\overline{)4}$

2. Here are 8 cookies. $\frac{1}{2}$ of the cookies are chocolate.

   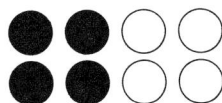

   How many cookies are chocolate?____

   Fill the blank: $\frac{1}{2}$ of 8 = ____

   Do this division on your paper: $2\overline{)8}$

3. Here are 8 marbles. $\frac{1}{4}$ of the marbles are black.

   How many marbles are black?____

   Fill the blank: $\frac{1}{4}$ of 8 = ____

   Do this division on your paper: $4\overline{)8}$

4. Here are 6 worms. $\frac{1}{3}$ of the worms are black.

   How many worms are black?____

   Fill in the blank: $\frac{1}{3}$ of 6 = ____

   Do this division on your paper: $3\overline{)6}$

122   FRACTIONS

5. Paul had 6 robots. He gave $\frac{1}{2}$ of them to Raoul.

   How many did he give to Raoul?____

6. A dragon had 10 bottles of gas. He drank $\frac{1}{2}$ of the bottles.

   How many bottles did he drink?____

7. Ten dinosaurs walked by a mountain. $\frac{1}{5}$ of them went in a cave.

   How many dinosaurs went in the cave?____

8. Twelve ants were eating a doughnut. $\frac{1}{3}$ of the ants got full and left.

   How many ants were left?____

9. Twelve ducks were swimming in a lake. $\frac{1}{4}$ of them got pulled under by the Lockneck Monster.

   How many ducks were pulled under?____

10. Twelve children were at a birthday party. $\frac{1}{6}$ of the children got sick from eating too much candy and had to go home.

    How many children got sick?____

Lesson 6

1. Carlos had 8 cookies. He gave $\frac{1}{4}$ of them to Sue.

   How many cookies did he give Sue?____
   Perform this division. 4)8

2. Tedra had 12 tennis balls. She lost $\frac{1}{4}$ of them.

   How many tennis balls did she lose?____

   Perform this division. 4)12

3. Fred has 12 jelly rolls. He ate $\frac{1}{3}$ of them.

   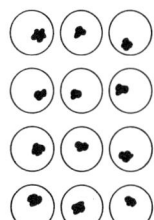

   How many rolls did he eat?____

   Perform this division. 3)12

4. Arlene has 12 bracelets. She gave $\frac{1}{6}$ of them to Kathy.

   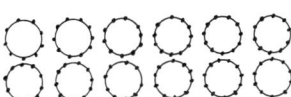

   How many bracelets did she give Kathy?____

   Perform this division. 6)12

5. Ken had 6 worms. He painted $\frac{1}{3}$ of them with red and white stripes to enter in a race.

   How many did he paint?____

   How many were not painted?____

124 FRACTIONS

6. Amanda had 20 peacock feathers. Her friend Louise borrowed $\frac{1}{5}$ of them to wear to a party.

   How many did Louise borrow?____

   How many did Amanda have left to wear to the party herself?____

7. To make swampsweet soup, Sally used $\frac{1}{4}$ of the 12 cabbage hearts she had in the refrigerator.

   How many cabbage hearts did she use?____

   How many cabbage hearts were left?____

8. Terry had 15 ants in his pants. He sat down and crushed $\frac{1}{3}$ of them.

   How many did he crush?____

   How many ants were still alive and crawling around in his pants?____

Lesson 7

1. A town had 15 buildings. A fire-breathing dragon burned down $\frac{1}{3}$ of the buildings.

   ⌂ ⌂ ⌂
   ⌂ ⌂ ⌂
   ⌂ ⌂ ⌂
   ⌂ ⌂ ⌂
   ⌂ ⌂ ⌂

   How many buildings did the dragon burn down.

   Do this division on your paper: 3)‾15‾

FRACTIONS 125

2. Fifteen termites were eating a house. $\frac{1}{5}$ of them decided to have a chair for lunch.

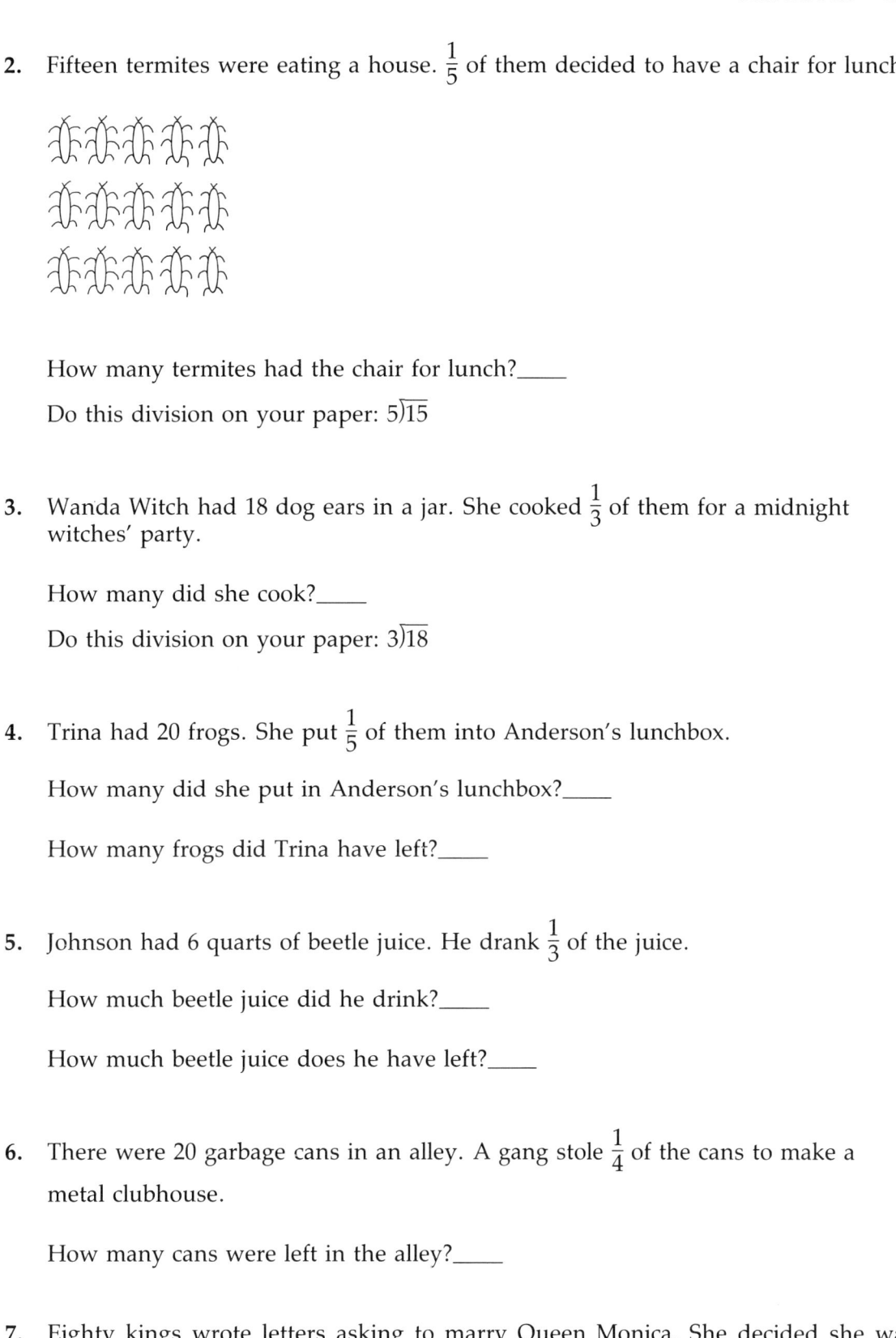

How many termites had the chair for lunch?____

Do this division on your paper: 5)15

3. Wanda Witch had 18 dog ears in a jar. She cooked $\frac{1}{3}$ of them for a midnight witches' party.

How many did she cook?____

Do this division on your paper: 3)18

4. Trina had 20 frogs. She put $\frac{1}{5}$ of them into Anderson's lunchbox.

How many did she put in Anderson's lunchbox?____

How many frogs did Trina have left?____

5. Johnson had 6 quarts of beetle juice. He drank $\frac{1}{3}$ of the juice.

How much beetle juice did he drink?____

How much beetle juice does he have left?____

6. There were 20 garbage cans in an alley. A gang stole $\frac{1}{4}$ of the cans to make a metal clubhouse.

How many cans were left in the alley?____

7. Eighty kings wrote letters asking to marry Queen Monica. She decided she was interested in $\frac{1}{4}$ of them. The rest were sent "get lost" letters.

How many were sent "get lost" letters?____

126 FRACTIONS

8. Marcus had 6 cans of nails and each can had 5 nails. He gave $\frac{1}{3}$ of the nails to Tina.

   How many nails did Marcus have left?____

Lesson 8

1. Tiffany had 30 video movies. She gave $\frac{1}{3}$ of them to her school.

   How many did she have left?____

2. Jungle Girl caught 15 bugs. She fed $\frac{1}{5}$ of them to her tigers. She cooked $\frac{1}{2}$ of the rest.

   How many bugs did she cook?____

3. Doris collected 17 cheesebars. She put 5 in the dog's food for extra calcium. She put $\frac{1}{6}$ of the rest in the cat's food.

   How many did she put in the cat's food?____

4. A store had 60 robots. One day they began attacking people. The people shot $\frac{1}{2}$ of them. But $\frac{1}{3}$ of the rest got away.

   How many robots got away?____

5. Early Bird caught 40 worms. He gave $\frac{1}{4}$ of them to his old Granny Bird. He ate $\frac{1}{5}$ of the rest on a salad.

   How many did he eat on his salad?____

6. Linda had 30 boxes of Girl Scout cookies. She sold $\frac{1}{3}$ of them on the weekend. She sold $\frac{1}{5}$ of the rest on Monday.

   How many did she sell on Monday?____

   How many boxes did she still have left?____

7. Michael had 50 baseball cards. He gave $\frac{1}{5}$ of the cards to Terrance. Then he gave $\frac{1}{5}$ of the remaining cards to Cassandra.

    How many cards did Michael have left?____

8. Johnson had 40 tennis balls. He lost $\frac{1}{5}$ of the balls. Then he found $\frac{1}{2}$ of the balls he had lost.

    How many tennis balls did Johnson have?____

Lesson 9

1. Judy found $6 on the street. She gave $\frac{1}{3}$ of the money to her sister.

    How much did she give her sister?____

    Do this division on your paper: 3)6

2. Mr. Gooddad won $40 in the lottery. He gave $\frac{1}{4}$ of the money to his son for his college fund.

    How much did he give his son?____

    Do this division on your paper: 4)40

3. A toy pink elephant normally sells for $8. During a sale, the toy store took $\frac{1}{4}$ off the price.

    How much did the store take off the price?____

    How much did the pink elephant cost on sale?____

4. A rabbit normally sells for $15. On sale the pet store took $\frac{1}{3}$ off the price.

    How much did the pet store take off the price?____

    How much did the rabbit cost on sale?____

5. A puppy normally sells for $60. On sale the pet store took $\frac{1}{6}$ off the price.

    How much did the puppy cost on sale?____

6. ●●○○○■□□□□

What fraction of the figures is black?____

Did you answer $\frac{3}{11}$ for the last question?

What fraction of the figures is white?____

What fraction of the circles is black?____

Did you answer $\frac{2}{5}$ for the last question?____

What fraction of the circles is white?____

What fraction of the squares is black?____

What fraction of the squares is white?____

Did you answer $\frac{5}{6}$ for the last question?____

7. Some children at a party drank shakes and others drank soda. Here are the children and what they drank.

   Paul—shake   Ruth—soda
   Tom—shake   Mary—shake
   Jerome—soda   Julie—shake
                 Suzie—shake

What fraction of the children had soda?____

Did you answer $\frac{2}{7}$ for the last question?

What fraction of the children had a shake?____

What fraction of the boys had soda?____

Did you answer $\frac{1}{3}$ for the last question?

What fraction of the boys had a shake?____

What fraction of the girls had a soda?____

What fraction of the girls had a shake?____

Did you answer $\frac{3}{4}$ for the last question?

8. There are 11 boys and 13 girls in a class. Seven of the boys and 8 of the girls like to watch cartoons.

    What fraction of the boys like to watch cartoons?____

    Did you answer $\frac{7}{11}$ for the last question?

    What fraction of the boys do not like to watch cartoons?____

    Did you answer $\frac{4}{11}$ for the last question?

    What fraction of the girls like to watch cartoons?____

    What fraction of the girls do not like to watch cartoons?____

    Did you answer $\frac{5}{13}$ for the last question?

    What fraction of all the children like cartoons?____

    What fraction of all the children do not like cartoons?____

    Did you reduce your answer of $\frac{9}{24}$ to $\frac{3}{8}$ for the last question?____

9. Alien had 5 spiders and 6 roaches. He ate 2 spiders.

    What fraction of the remaining bugs are roaches?____

10. Harriet bought 4 cheeseburgers and 9 bacon burgers. She gobbled down 1 cheeseburger and 2 bacon burgers while driving home.

    What fraction of the remaining burgers are cheeseburgers?____

11. A room where a movie is being shown has 24 seats. The room was $\frac{1}{2}$ full. But $\frac{1}{3}$ of the people got bored and left.

    How many people are still watching the movie?____

Lesson 10

1. Copy this picture on your paper.

   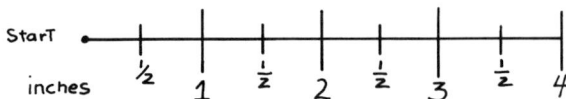

   A baby ant can crawl $\frac{1}{2}$ of an inch each minute. It begins at the START.

   Write A on your picture where the ant will be after 1 minute.

   Write B where the ant will be after 3 minutes.

   How far from the start is the ant after 3 minutes?____

   Did you answer $1\frac{1}{2}$ inches for the last question?

   Write C where the ant will be after 4 minutes.

   How far from the start is the ant after 4 minutes?____

   Write D where the ant will be after 7 minutes.

   How far from the start is the ant after 7 minutes?____

2. This is a picture of a race track which has a mark every $\frac{1}{4}$ of a mile so racers can tell how far they have gone.

   Copy this picture.

   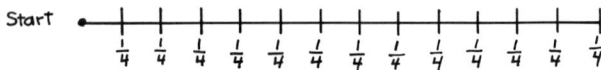

   In a race, John's wheelchair travels $\frac{1}{4}$ mile in 1 minute. He begins at the START.

   Write A where John is after 3 minutes.

   How far from the start is John after 3 minutes?____

   Did you answer $\frac{3}{4}$ mile for the last question?

   Write B where John is after 5 minutes.

   How far from the start is John after 5 minutes?____

   Write C where John is after 10 minutes.

   How far from the start is John after 10 minutes?____

   Did you answer $2\frac{1}{2}$ miles for the last question?

3. This is a glass that is shaded to look $\frac{1}{3}$ full.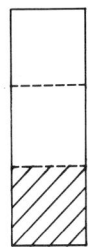

Draw this glass and shade it to look $\frac{2}{3}$ full.

4. This is a glass shaded to look $\frac{2}{5}$ full.

Draw this glass and shade it to look $\frac{4}{5}$ full.

132 FRACTIONS

5. Here are 2 glasses shaded to look $1\frac{1}{4}$ full.

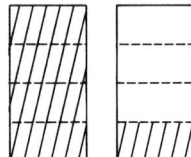

Draw these 2 glasses and shade them to look $1\frac{3}{4}$ full.

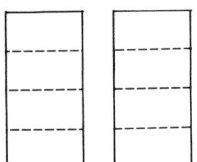

Draw these 3 glasses and shade them to look $2\frac{1}{4}$ full.

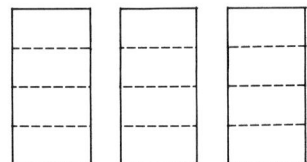

6. You use $\frac{1}{3}$ cup of mayonnaise to make 1 serving of tuna salad.

⅓ cup

Copy this table.

Indicate how many cups of mayonnaise are needed for each number of servings of tuna listed.

Servings of Tuna	Cups of Mayonnaise
1	_____
2	_____
3	_____
4	_____
5	_____ answer $1\frac{1}{3}$ →
6	_____
7	_____
9	_____
11	_____
13	_____

7. Paul had $1\frac{1}{5}$ cups of red paint and Judy had $2\frac{3}{5}$ cups of red paint.

How much paint did they have together? _____

8. Dracula had $1\frac{2}{7}$ cups of blood and Gordo had $\frac{3}{7}$ cups of blood.

   How much blood did they have together?____

9. Dracula had $1\frac{5}{7}$ cups of blood. He drank $\frac{3}{7}$ cups.

   How much did he have left?____

Lesson 11

1. The cafeteria serves pie for dessert. Each serving is $\frac{1}{5}$ of a pie.

   How many servings can be made from 1 pie?____

   How many servings can be made from 3 pies?____

   How many pies do you need to make 10 servings?____

2. Here is a cake cut into 4 equal pieces.

J	J
J	S

   John ate the pieces marked with a J.

   What fraction of the cake did John eat?____
   Sammy ate the pieces marked with an S.

   What fraction of the cake did Sammy eat?____

   Do this addition on your paper: $\frac{3}{4}$ cake + $\frac{1}{4}$ cake = ____ cake

134 FRACTIONS

3. Draw these two cakes. Include the dotted lines showing how each cake can be cut into 5 equal pieces.

Mario ate $\frac{3}{5}$ of a cake.

Write M in the pieces Mario ate.

Jeremy ate $1\frac{1}{5}$ cake.

Write J in the pieces Jeremy ate.

How much cake did the two boys eat?____

Do this addition on your paper. $\frac{3}{5} + 1\frac{1}{5} =$ ____

4. Here are 2 chocolate bars each cut into 5 equal pieces.

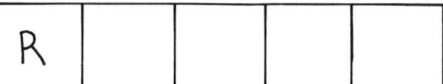

Theresa ate the pieces marked with T.

What fraction of a chocolate bar did she eat?____
Rolando ate the pieces marked with R.

What fraction of a bar did Rolando eat?____

How much chocolate did the children eat together?____

Do this addition on your paper: $\frac{2}{5} + \frac{4}{5} =$ ____

5. Here is a square pizza cut into 4 equal pieces.

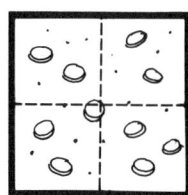

Lucia ate $\frac{1}{4}$ of the whole pizza.

How much was left?____

Then Frank ate $\frac{1}{4}$ of the whole pizza.

Now how much is left?____

Do this subtraction on your paper: $1 - \frac{1}{4} =$ ____

Do this subtraction on your paper: $\frac{3}{4} - \frac{1}{4} =$ ____

Do this addition on your paper: $\frac{1}{2} + \frac{1}{4} =$ ____

Do this addition on your paper: $\frac{3}{4} + \frac{1}{4} =$ ____

6. There were 2 peanut blocks each divided into 3 equal pieces on the table.

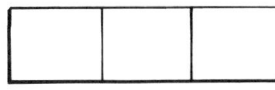

Danny's pet bear ate $\frac{1}{3}$ of a block.

How much was left?____

Then Danny ate $\frac{1}{3}$ of a block.

Now how much is left?____

Do this subtraction on your paper: $2 - \frac{1}{3} =$ ____

Do this subtraction on your paper: $1\frac{2}{3} - \frac{1}{3} =$ ____

Do this addition on your paper: $1\frac{1}{3} + \frac{1}{3} =$ ____

Do this addition on your paper: $1\frac{2}{3} + \frac{1}{3} =$ ____

Lesson 12

1. There was $\frac{2}{3}$ of a quart of ice cream in the refrigerator.

   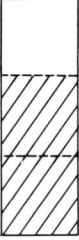

   Lamar ate $\frac{1}{3}$ of a quart of ice cream.

   How much ice cream was left?_____

   Do this subtraction on your paper: $\frac{2}{3} - \frac{1}{3} = $ _____

2. Darren had $\frac{5}{7}$ of a gallon of gas in his motorcycle. He used $\frac{3}{7}$ of a gallon going to a picnic.

   | 1 | 2 | 3 | 4 | 5 | 6 | 7 |

   How much gas did he have left?_____

3. Laura bought 2 rolls of orange paper to make a pumpkin costume for Halloween. She used $1\frac{1}{3}$ rolls.

   How much was left over?_____

4. Frank bought 8 rolls of blue paper to make a super hero costume. He used $3\frac{1}{4}$ rolls to make the suit and 2 rolls to make a cape.

   How much was left over?_____

   Do this addition on your paper: $3\frac{1}{4} + 2 = $ _____

   Do this subtraction on your paper: $8 - 5\frac{1}{4} = $ _____

5. Mailman Manny's mailbag weighed $11\frac{4}{5}$ pounds. He delivered one package weighing $2\frac{1}{5}$ pounds and another weighing $4\frac{2}{5}$ pounds.

   Then how much did his mailbag weigh? ____

   Do this addition on your paper: $2\frac{1}{5} + 4\frac{2}{5} =$ ____

6. It takes Nick $\frac{1}{4}$ of an hour to walk to school.

   How much time does Nick spend walking to school and then back home each day? ____

   How much time does Nick spend walking to school in a 5-day school week? ____

   How much time does Nick spend walking to school in a month with 20 school days? ____

7. Four children had lunch. Each took $\frac{1}{2}$ of an orange in the cafeteria.

   How many oranges did they have altogether? ____

8. Ten people had lunch. Each had $\frac{1}{2}$ of a melon.

   How many melons did they have altogether? ____

9. Four people had lunch. Each has $\frac{1}{3}$ of a pie.

   How much pie did they eat altogether? ____

10. Karen is $5\frac{1}{2}$ years old. Her cat Tom is 2 years old.

    How much older is Karen than Tom? ____

11. Pam had 5 cans of dog food. She feeds her dog Toodles $\frac{3}{4}$ of a can each day.

    How much is left after feeding Toodles 1 day? ____

    How much is left after 2 days? ____

    How much is left after 5 days? ____

Lesson 13

1. Erron had $\frac{1}{2}$ of a pie. Cliff gave him $\frac{1}{4}$ of a pie.

   Then how much pie did Erron have?____

   Do this addition on your paper: $\frac{2}{4} + \frac{1}{4} =$ ____

2. Helen had $\frac{1}{2}$ of a pizza. Her mother gave her $\frac{1}{3}$ of a pizza.

   Then how much of a pizza did she have?____

   Do this addition on your paper: $\frac{3}{6} + \frac{2}{6} =$ ____

3. Jill bought $\frac{1}{2}$ of a cake. She ate $\frac{1}{3}$ of a cake.

   How much did she have left?____

   Do this subtraction on your paper: $\frac{3}{6} - \frac{2}{6} =$ ____

4. Bob had $\frac{1}{5}$ of a gallon of gas in his lawn mower. He added $\frac{1}{2}$ of a gallon.

   Then how much gas did he have?____

5. Bob had $\frac{7}{10}$ of a gallon of gas in his lawn mower. He used $\frac{1}{2}$ of a gallon mowing his lawn.

   How much gas did he have left?____

6. Denise bought a box of candy weighing $1\frac{1}{2}$ pounds. She ate $\frac{1}{3}$ of a pound.

   How much candy was left?____

FRACTIONS 139

7. Emmanuel had $\frac{1}{2}$ quart of beetle juice. Jenny had $3\frac{1}{2}$ quarts of beetle juice. They put all the juice together in one jar, but they spilled $1\frac{1}{2}$ quarts.

   How much juice was left in the jar?____

8. Four quarts equal 1 gallon.
   Sarah has $3\frac{1}{2}$ quarts of punch.

   How much more does she need to fill a gallon bowl?____

9. Four quarts equal 1 gallon.
   Frank caught baby sharks that must be kept in salt water. He has $2\frac{1}{4}$ quarts of salt water.

   How much more does he need to fill a 1 gallon fish tank?____

10. Mr. Dricar spends $\frac{1}{4}$ of a dollar every day for a bridge toll.
    How much does he spend in 3 days?
    Write your answer as a fraction of a dollar.

    fraction answer:____

    Write your answer in cents.

    cents answer:____

    How much does he spend in a 5-day work week?____

    dollars with fractions:____

    dollars and cents:____

11. Candy worms cost $\frac{1}{10}$ of a dollar each.
    How much did Karen spend for 5 candy worms?
    Write your answer as a fraction of a dollar.

    fraction answer:____

    Write your answer in cents.

    cents answer:____

FRACTIONS

12. Jelly apples cost $\frac{1}{2}$ of a dollar each.

How much did Judy spend for 3 jelly apples?____

dollars with fractions answer:____

dollars and cents answer:____

How much did Regina spend for 6 jelly apples?____

4 Fractions: Additional Exercises

Lesson 1x

1. Tanya found $\frac{1}{3}$ of a torn dollar bill. Jackie found another $\frac{1}{3}$ of the same bill.

   How much of the bill do they together have?____

2. Brian is 10 years old. His brother Juan is $1\frac{1}{2}$ years older. His sister Anna is $2\frac{1}{2}$ years older than Juan.

   How old is Anna?____

3. Marge picked $\frac{3}{5}$ of a pail of berries. She ate $\frac{1}{5}$ of a pail.

   How much was left?____

4. Cindy runs $10\frac{1}{2}$ miles a week. Chris runs 6 miles a week.

   How much more does Cindy run than Chris?____

5. Albert is 13 years old. His sister Cindy is $3\frac{1}{2}$ years older. His brother John is 5 years younger than Cindy.

   How old is John?____

6. Charles cooked dinner for himself, his 3 sisters, his brother, and his father. He made a $\frac{1}{4}$ pound hamburger for each person.

   How much meat did he use?____

7. Gloria bought a watch for $\frac{1}{4}$ off the regular price of $36.

   How much did she pay for the watch?____

142 FRACTIONS: ADDITIONAL EXERCISES

8. Claud wants to make an omelet with $\frac{1}{3}$ of a dozen eggs. He only has 1 egg.

   How many more eggs does he need?____

9. Glen bought 1 dozen eggs. On the way home he tripped and broke $\frac{1}{3}$ of them.

   How many eggs were broken?____

   How many eggs were not broken?____

10. Carmen bought 48 baseball cards. She gave $\frac{1}{3}$ of them to her cousin. Then she gave $\frac{1}{4}$ of the remainder to her sister.

    How many did she give her sister?____

    How many cards did Carmen have left?____